Crossing Places
of the
Upper
Thames
A History & Guide

CROSSING PLACES OF THE UPPER THAMES
A HISTORY & GUIDE

AMY WOOLACOTT

To Amy Sabina Blackmore – the grandmother I never knew.

Line drawings by Katherine Blackmore and photographs mostly by the author. The author claims the copyright to the contents except where otherwise stated.

First published 2008
Reprinted 2019

Tempus Publishing, The History Press Ltd
97 St George's Place, Cheltenham,
Gloucestershire, GL50 3QB
www.thehistorypress.co.uk

Tempus Publishing is an imprint of The History Press Ltd

© Amy Woolacott, 2008

The right of Amy Woolacott to be identified as the Author of this work has been asserted in accordance with the Copyrights, Designs and Patents Act 1988.

All rights reserved. No part of this book may be reprinted or reproduced or utilised in any form or by any electronic, mechanical or other means, now known or hereafter invented, including photocopying and recording, or in any information storage or retrieval system, without the permission in writing from the Publishers.

British Library Cataloguing in Publication Data.
A catalogue record for this book is available from the British Library.

ISBN 978 0 7524 4693 6

Typesetting and origination by The History Press Ltd
Printed in Great Britain by TJ International Ltd, Padstow, Cornwall.

Contents

		Acknowledgements	7
		Sources	9
		Introduction	11
	one	The Evolution of River Crossings	15
	two	Navigation Improvements	23
	three	The Future	35

Part Two

Section one	Source to Cricklade	39
Section two	Cricklade to Lechlade	67
Section three	Lechlade to Newbridge	89
Section four	Newbridge to Oxford	115

'Father' Thames beside St John's Lock, Lechlade.

Acknowledgements

I am grateful to many of my friends who prefer to remain anonymous; their interest, encouragement and companionship have been so fortifying. I would like to acknowledge help from the late Geoffrey Beddow who gave much useful early guidance, and the good advice offered by the Gloucestershire author Gordon Ottewell. I would also like to thank Margaret and Eric Miller for proofreading and for their helpful suggestions. I have long hesitated to launch this work and expose it and myself to criticism from professional historians.

Sources

In compiling this book countless local guidebooks and histories, frequently found in town and village churches, were great sources of local knowledge. Reference works from the English Place-Name Society for the counties through which the Upper Thames flows give details of early forms of place names; *Phillimore Domesday Book* volumes identify pre-Conquest mills and fisheries; and *Victoria County Histories* (OUP) for the four counties around the Upper Thames give town and village histories.

Much additional historical and geographical detail was gleaned from a wide variety of maps: Enclosure and Tithe Awards; Richard Davis's map (1790s) of Oxfordshire engraved by John Cary; surveys of Oxfordshire and Berkshire by John Rocque in the mid-1700s; Isaac Taylor's Gloucestershire; Bowden's 1775 map; Greenwood's 1824 map; and First Edition Ordnance Survey (OS) maps in the nineteenth century. The works of early cartographers – Saxton, Speed, Morden, Moule and Ogilby – were funds of information, as were the twentieth-century OS 1in series, Landranger (1:50,000) and Pathfinder (1:25,000) Metric Series, and the OS 2½in street atlases of Oxfordshire, Berkshire and Gloucestershire.

Books that gave inspiration during writing:

J.R.L Anderson, *The Upper Thames* (1970)
Walter Armstrong, *The Thames from its Rise to The Nore* (1889)
John Blair, *Anglo-Saxon Oxfordshire* (1994)
H.W. John Cuss, *The Valley of the Upper Thames* (1998)
Daniel Defoe, *A Tour through the Whole Island of Great Britain* [1714–26] (1971)
Brian Eade, *Along the Thames* (1997)
Margaret Gelling, *The Place Names of Berkshire* – three volumes (English Place-Name Society, 1974)
Margaret Gelling, *The Place Names of Oxfordshire* – two volumes (English Place-Name Society, 1971)
J.E.B. Gover, A. Mawer & F.M. Stenton, *The Place Names of Wiltshire* (English Place-Name Society, 1992)
Carolyn Heighway, *Anglo-Saxon Gloucestershire* (1987)
Humphrey Household, *Gloucestershire Railways in the Twenties* (1986)
Humphrey Household, *The Thames & Severn Canal* (1969)
A. Jenkins, *The Book of the Thames* (1983)
John Leland, *Itinerary* [1535–43] (1964)
David Perrott (editor), *Ordnance Survey Guide to the River Thames* (1994)
Geoffrey Phillips, *Thames Crossings* (1981)
Anne Savage (Translation and Collation of) *Anglo-Saxon Chronicles* (1996)
Alan Saville (editor), *Archaeology in Gloucestershire* (1984)
David Sharp, *The Thames Path* (1996)
A.H. Smith, *The Place Names of Gloucestershire* – four volumes (English Place-Name Society, 1964–65)
John Steane, *Oxfordshire* (1996)
F.S. Thacker, *The Thames and its Story* (1906)
F.S. Thacker, *The Stripling Thames* (1909)
F.S. Thacker, *The Thames Highway Volume I – General History* (1914)
F.S. Thacker, *The Thames Highway Volume II – Locks & Weirs* (1920)
A. Williams, *Round about the Upper Thames* (1922)
D.G. Wilson, *The Victorian Thames* (1993)

Introduction

The Upper Thames was once much wider and shallower than it appears today, with numerous side channels that formed during winter, but perhaps dried out in summer. A river shapes a landscape, fosters plants and wildlife, but it can be a highway too, or serve as a boundary between territories. When a land bridge between Britain and Europe was still open primitive humans moved northwards. Doubtless early settlers used some sort of boat, perhaps a dugout canoe or hide-covered craft, enabling pioneers to explore new territory. After the final retreat of the great ice-sheet from Britain, migrants found that the Upper Thames Valley supported life of all kinds; settlers used riverside land as a valuable resource for stock grazing and hay crops and for rushes and osiers. They managed the river to harvest its fish and harnessed its power to drive mills. Over the centuries settlements grew with flourishing trade that extended far beyond their bounds, leading to many cross-country and cross-river routes being formed.

My intention was to shed light on the history of the Upper Thames crossings, and relate their origins to tracks serving local communities, though these were likely connecting with long-distance trade routes. Many happy hours were spent looking for hints of history: walking or cycling riverside footpaths, or travelling by dinghy to visualise how our ancestors interpreted the landscape. My thoughts constantly turned to early usage of the river as a highway. Poets have called the Thames stately and silvery, and it certainly adds charm to the villages and towns along its course. While taking sustenance in village inns there were often chance meetings with local people able to share their knowledge of the river and the surrounding countryside. These experiences suggest that travelling at a gentle pace, like our forefathers did, makes for a deeper understanding of those bygone days, enhanced by enjoyment of the countryside of today; an escape from the modern world, a freedom to enjoy nature denied to those travelling in fast cars.

In my search for older crossing locations I often deplored some of the decay that has occurred, but I welcomed recent improvements and occasionally witnessed history in the making. The various types of crossing are shown with line drawings, or photographs of the more significant crossings along the Upper Thames. Four maps show the general trend of the river and the towns or villages a short distance from it using a river crossing. The settlements that contributed to the history of cross-river communication might also be useful centres for today's visitors. More than 150 crossing places of the past and present were found. Their locations are identified with OS grid references within the text, and whatever their former uses or importance today these crossing points are like windows of time along a timeless river. Accurate information on the continued existence of public rights-of-way will be found in the most recent Ordnance Survey maps.

Upper Thames Settlements

The Upper Thames is roughly a quarter of the river's 210-mile course. Alluvial deposits in the valley often reveal traces of early human habitation principally with animal bones and flint tools such as Palaeolithic hand-axes, which may be more than 200,000 years old. Because they are not of local origin, they no doubt arrived with nomadic travellers or traders. Migrant hunter-gatherers later moved through the landscape making only flimsy shelters, but it may be claimed with fair certainty that the first permanent buildings were where prehistoric settlers began farming. Much later there were Romanised centres, occupied for indeterminate periods, where Iron Age farmers had previously settled.

The Saxons came to the Upper Thames plains by the late fifth century, and many believe that Saxon villages grew where Iron Age farmers previously tilled the arable soil or benefited from the good grazing land. It is almost certain that the Romano-British were not ousted by the Anglo-Saxon conquerors, who increased land usage and chose to settle near streams and rivers. By the seventh century West Saxons had large open-field farms using marshland for hay meadows drained by newly cut ditches. They also created numerous mills and fisheries, and exploited the resources of nearby woodland. Many rivers have the oldest names in the landscape and some pre-date the arrival of the Romans. Old English for river was *ea*, which occurs at Eaton settlements. Many Thames-side places such as Osney, Hinksey, Chimney and Rushey use the Old English name element *ieg* meaning 'land amid marsh'. The term *moor*, occurring at Northmoor and Southmoor, described grassland in floodplain marshes.

Whilst it is true that most of our forebears led mundane lives, many contributed to the colourful history we find in the towns and villages along the river valley, suggesting they were ideal places for habitation. Meanwhile, 'Father' Thames continually adds to history and provides recreation.

one

The Evolution of Crossing Places

Permanent river crossings would have been on routes important enough to justify their construction, and many ancient crossing places are still in use. Some crossings leave subtle hints of their former economic or military importance. Many modern routes began in our Saxon era; Anglo-Saxon charters mention river crossings more than 600 times and over 80 per cent of them were fords. Where the charters mention bridges, the implication is they were Roman bridges of timber or stone. With growing early medieval commerce, new bridges brought conflict with users of the Thames for navigation. Which was more important to travellers and traders – the roads or the river?

Fords

The Upper Thames in its natural state offered numerous fording places, many of which remained in use long after an adjacent bridge was built. It is known that the Romans laid fords across the Thames both east and west of Oxford. Fords varied in depth and some clues are provided by place names. Shifford (sheep-ford) suggests it was shallow enough for the short-legged woolly animals, whereas Swinford (ford of swine) implies one a little deeper. At Oxford the ford for longer-legged cattle would have been much deeper. Pedestrians could have crossed downstream of fish-weirs that often created convenient shallows, but weirs and fords would have hindered larger river-vessel movements.

An early term for ford was *gewaed*, used in Anglo-Saxon boundary descriptions. The term *gelade* probably described a way across water where land was likely to flood. In AD 895 *gelade* was used in land charters and in the mid-tenth century *ford* occurred in descriptions of many estates along the Upper Thames. A charter of 956 has a clause identifying a *fordwere* (ford-weir).

Early wayfarers frequently waded through water up to their thighs, though some fords were described as *sceald* (shoaly water). Until the nineteenth century, fords were often nearly a metre deep, but now only a few ancient, and mercifully shallow, fords remain on the Upper Thames. A 940s document describes fords with (*gyrd*) stakes or (*stapol*) pillars built into the stone foundations, possibly as guides (maybe with a rope strung between them); this suggests that travellers needed a degree of confidence to undertake a crossing. About that time a *stan* or stone-ford was mentioned, implying stone laid to improve the crossing, possibly done originally by the Romans; we might even call this a causeway. A land charter of 956 refers to a *brycg ford*, where the Old English word *brycg* meant a causeway. The term *brycg* probably originated in the early eighth century.

Bridges

Bridges are often inspiring, instantly recognisable and even world famous, a good example being London's Tower Bridge built in 1894. Britain's prehistoric inhabitants

Ford with footbridge.

A Clapper bridge on the River Leach.

constructed bridges from suitable locally available materials. Timber could have been used to create artificial islands to help span a wide part of the river. Large flat stones used as stepping-stones are identified in some Anglo-Saxon land charters, though none occur on the Upper Thames. The logical successor to stepping-stones was a clapper bridge over a shallow stream, a modern example being at Eastleach on a Thames tributary, the river Leach. A charter of 923 identifies a *wudubricge* (wooden bridge), and in another of 958 a *stanbricgge* (stone-bridge) on the Thames is mentioned.

New bridges were frequently built near existing fords and not over a narrower part of the river, which suggests that routes were already well established. A modern clue to a previous fording place will show where a road widens on approaching a bridge, hinting that the bridge and ford co-existed for some years. River-crossings are referred to in pre-Conquest land charters on hundreds of occasions, though fewer than 20 per cent were bridges, but as early as 749 we know that Aethelbald of Mercia charged monastic communities with bridge repairs. The duty of bridge building and repair was later termed essential work from which no one was excused.

The medieval era marked a revival of urban life. The Romans withdrew from Britain in the fifth century, but their stone bridges hardly changed until monks revived bridge-building skills in the eleventh century. The Roman writer Vitruvius recorded details of bridge construction in the first century BC and early medieval builders no doubt followed similar procedures. Some landowners employed the local militia to build bridges, which became important to traders, and the number of bridges increased from the mid-thirteenth century onwards. King John was an avid traveller around his kingdom, which might account for the many new bridges constructed during his reign. Many timber bridges were rebuilt in stone with approach causeways, but all too often a splendid medieval bridge linked tracks that were already deteriorating with increased use. Most causeways (or causeys) were built of deeply embedded stone, but were dismantled later when better roads and improved land drainage began from the seventeenth century onwards.

The bulk of travellers were merchants with waggons or packhorses. Packhorse bridges, first appearing in the fourteenth century, were usually narrow and with low parapets so that bundles of merchandise were not jostled. Other frequent travellers were bishops, preachers, pilgrims and taxmen. With the Dissolution of the Monasteries there came a decline in long-distance travel; few new bridges were built between the late 1500s and the early Georgian era.

From the late thirteenth century the lord of each manor was obliged to maintain bridges and roads, and had the right to impose tolls (pontage) for passage over (and under) bridges, which began when the construction was completed. Many local bailiffs collected the tolls. In times of war, bridges became key places for controlling navigation and militia movements. Stone bridges, though costly to construct, were doubtless demanded for safer crossings on busy routes, and were frequently funded from charitable bequests or merchants' guilds. However, after 1547, endowments from guilds were suppressed, and most ceased altogether after the Reformation, but local Justices of the Peace or landowners became responsible for building most new bridges and their maintenance.

Bridge builders were master masons and also architects. By and large, they were itinerant journeymen until the late sixteenth century. St Peter became patron saint of medieval bridge craftsmen, who also had their own local saints. Masons used designs that their brothers used to shape arches in cathedrals and churches, but added special bridge features such as cutwaters that gradually evolved to the efficient pointed form. Modifications to old bridges may be visible as an increase in width or reshaping of arches for the benefit of river traffic.

Bridge architecture gradually became more elaborate, sometimes with a fortified gate. The Monnow Bridge at Monmouth is a surviving example of this. New bridges of the eighteenth century were often of Renaissance design. Many incorporated houses and shops; the old London Bridge was typical of this, carrying the only City road across the river until 1783. Pulteney Bridge (1769) on the Avon at Bath is a surviving example with shops. The Victorians reintroduced Gothic styling in stone, and then in 1779 the world's first iron bridge was constructed across the Severn at Coalbrookdale, now known as the Ironbridge Gorge. Cast-iron bridges became commonplace in the nineteenth century but iron and steel need regular maintenance, whereas good quality stone is almost imperishable. Most modern bridges are of reinforced or of pre-stressed concrete, though carbon-fibre has been used in experimental situations. Surprisingly, England's last timber road-bridge at Selby, Yorkshire, was not demolished until the 1970s.

Ferries

Seasonal changes in depth of water at fords brought increased demand for ferries, but even these experienced hazards that bridges mostly eliminated. In Domesday Book, *Feribi* (from Old Norse place-name elements *ferja* and *by*) occurs for North and South Ferriby on opposite banks of the River Humber, identifying a ferry route created late in our Danish era. The terms *ferry* and *ferryman* came into more general use by the twelfth century, initially to signify transporting people and goods from place to place. Specific use to describe working a craft across water, usually for a 'fare', came a little later.

In the thirteenth century hermits were hired for duty at fords and ferries. Ferry tolls were a steady source of income and, because boat maintenance costs were relatively low, many ferries continued in use beside newly built bridges often with a higher toll. In the early nineteenth century there was growing hostility to ferries from those using the river for trade – probably because many ferries used a rope or chain between the banks. Most of the few ferries on the Upper Thames were discontinued during the Second World War.

A barge ferrying livestock.

Highway Improvements

The old Roman highways were long overdue for improvement. Old routes were used increasingly for access to market towns, or to towns where rivers formed part of the transport network. Parishioners were obliged to contribute to the maintenance of roads. Bridge maintenance improved the crossings but bridges often brought conflict with river traders. The Turnpike Act (1663) allowed for road-tolls to fund improvements, and the term 'turnpike' came into more general use by 1678. The General Turnpike Act (1773) required other improvements such as mileposts, and some bridges were lengthened to eliminate their 'hump'. Nineteenth-century Highways Boards created non-turnpike roads with rates levied on landowners. In about 1886 County Councils took responsibility for upkeep of most roads and bridges.

Travel Hazards

Travel in early times was seldom free from hazards and those embarking on a journey would often invoke spiritual protection. Archaeologists say that in prehistoric times river 'gods' were often placated with carved stone figures of pagan deities, only to be discovered later within old bridge foundations. Those whose journey included a river crossing believed that St Julian Hospitaller, a mystic depicted in a boat or carrying an oar, protected the ferry and ferryman. Those in fear of drowning summoned St Romanus, a fifth-century hermit, or St John of Nepomuk, patron saint of Bohemia who was martyred in 1393. He was tortured and bound in chains before being thrown from a bridge into the River Vltava. Medieval folk believed that a holy cross or saint's statuette mounted on a bridge divinely protected the structure. Some bridges had a little chapel; perhaps where the founder's soul would be prayed for, or where a hermit might collect tolls towards bridge repairs. Bradford-on-Avon Bridge chapel is a good example dating from 1397, though it later became the town lock-up.

Many journeys would have been to markets, or to country fairs celebrating local saints' feast-days, and would have been particularly well patronised if held in summer when food was abundant. Saint cults and belief in healing springs were very strong in the Middle Ages but had much older origins, and numerous saints' shrines were often enriched by royal patronage. St Anthony was popular with medieval travellers and was the patron saint of packhorse traders. Merchants, fishermen and sailors claimed the patronage of St Nicholas of Myra. St Christopher is known universally as a patron saint of travellers; indeed the belief is so strong that space-voyagers in Apollo-8 took an amulet with them. The 'holy man' is frequently depicted in medieval church murals, and kindly hermits often helped wayfarers to cross over rivers.

two

Navigation Improvements

Water transport was always a very economic method of moving goods, and even the shallowest of rivers, which today might be sluggish ditches, were used for transport even before Roman times. The Thames is a sizeable river of the southern Midlands and the Saxons regulated its flow with weirs, partly to benefit millers or fishermen, but all too often the weirs interfered with navigation. Medieval millers and fishermen were major opponents to navigation, perhaps because of the 1650s by-laws that stopped mills grinding when the river flash-weirs were open. Previously, boats were delayed for several days because mill-owners did not want the water levels jeopardised. Some millers even claimed that wells sunk in towns and villages adversely affected river headwaters.

River management means making adjustments to the river's natural course and controlling water levels to ease navigation. It is not known who first instigated navigation improvements of the Upper Thames, but early in the eleventh century the Benedictine monks of Abingdon Abbey cut a navigation channel known as Swift Ditch, south of Andersey Island at Abingdon. Leland writing in 1535 says the channel was the chief stream. The abbey, originally founded in 675 but destroyed by the Danes around 790, was re-established by the black-robed monks around 960 under Abbot Aethelwold. He had the new channel cut to purge the abbey latrines. This brought a Thames channel closer to the town. The abbot owned extensive estates, farms and watermills on the river near Newbridge only six miles overland to the west. By the fifteenth century local merchants had paid for a stone bridge at Abingdon to improve goods movements by road.

Nevertheless, the Thames remained important for transport of bulky, heavy goods, downstream particularly, because it cost about 60 per cent less than transport by road. Historians tell us that the Thames was open as far as Oxford in the early thirteenth century, but in around 1350 it was required that the river should be clear up to Radcot, which at that time was said to be the navigation terminus. During the reign of Edward III (1327–77) parts of the Thames were widened so that barges could reach Oxford without hindrance.

Later improvements to the river, with some stretches being effectively canalised in the sixteenth century, led to poorer access to some smaller tributaries. The first Act of Parliament to improve navigation upstream of Oxford was passed in 1605, but progress was held up by the Civil War. Very soon horse-drawn boats travelled westwards from Oxford. Their tonnage rose to 60 tons in the mid-1700s and consequently greater depths of water were required. In the mid-nineteenth century river travel was still cheaper than by rail, but not nearly as speedy.

Mills and Weirs

The Upper Thames was once wide spreading with large lagoons and many parallel channels due to its shallow gradient. Even before the Norman Conquest, it drove hundreds of mills; indeed, almost every riverside parish would have had at least one because they were the mainstay of rural life. Mills, though expensive to build, raised valuable revenue from grain milling. The Romans introduced watermills to England in the second century, and though the earliest leats were Roman the term 'leat' for a man-made channel was actually coined in Saxon times, using Old English *gelaet*. Previously, the term was 'cut', and there was a Cutmill at Stanton Harcourt, near Eynsham, until the 1270s.

A mill-leat, controlled by a weir, could be over a mile in length, and invariably had at least one small footbridge. Leats sometimes included eel-traps, as shown in a margin drawing of the Luttrell Psalter (*c.*1335). Nevertheless, in the same era Chaucer was lamenting the demise of waterwheels and their weirs. Milling benefited a wide variety of rural industries such as cloth-fulling, paper-making, bonemeal-grinding for fertilizer and chaff-cutting, and some farmers used a mill tailrace as a sheep-wash.

Anglo-Saxon descriptions of monastic or royal land often marked boundary points at weirs or watermills and some can be identified as existing today, though little of the original structures will have survived. The Domesday Survey mentions thousands of mills on sites so well chosen that they were perhaps already over 100 years old. Few new watermills were built for many years (windmills were unknown in Britain before about 1150). Some early millwheels were set horizontally in the water directly below the millstones; others had the vertical wheels we see today, though these were sometimes altered to a horizontal setting.

Weirs appear in records as early as the 570s and attempts to control the Thames are known to historians from many pre-Conquest land charters. One dated to 957 refers to 'a weir at a nook of the Thames', though in Saxon times a *wera* was a wickerwork fish-trap. Monastic communities built weirs for *fiscere* (fishermen), which created fish-pools (*fisc-pol*), but these weirs often did not span the whole river. Weirs are shown on old maps with various spellings, from *were* or *waer* to *whire* or *weyr*, and in later times they are often prefixed with a weir-keeper's name. Some weir-keepers kept an adjacent inn or operated a ferry – both traditional activities – and they collected tolls from wayfarers and fishermen for centuries, though tolls were very often paid in goods such as eels or grain.

Weirs helped reduce unwanted shallows; this eased navigation, but in times of flooding riverside dwellers complained that crops were ruined when rivers over-spilled their banks. However, mills and weirs actually slowed the river's flow, ultimately causing it to silt up. Weirs became a hindrance to larger boats and in the late sixteenth century demands were still being made for their removal.

Flash-weirs

Flash-weirs eased the passage of river-craft and regulated river flow, but they also made a dry-shod crossing possible because the construction was a bridge of beams supported on posts (*rymers*) set into a socketed beam laid in the riverbed. Boards (*paddles*) were lodged between the posts to stem the river's flow, but could be lifted for the passage of boats. Unfortunately, the 'flash' of water released when the 'paddle-and-rymer' weir was opened meant the river level above was often depleted for several hours. This delayed other traffic and sometimes affected the operation of mills located nearby. Boats travelling upstream would have to be hauled through open weirs on stout ropes, perhaps using a capstan mounted on a

Form and function of a flash-weir.

well-anchored ground-frame on the towpath. A few paddle-and-rymer weirs have survived adjacent to some modern locks. It was common for flash-weir keepers to care for more than one weir, and it was their duty to trim weir levels and conduct flood control. Their knowledge and skills were handed down through their families.

Weirs of stakes, wattle or brushwood and small boulders along the base often had the stout timbers clad with lead to prevent premature decay. In Henry VIII's time a weir (on the Avon at Ringwood, Hampshire) had eighteen floodgates. Each gate was 2ft 6in wide, with every post set arris-wise, that is, with the sharp edge set towards the flow. The shallow water below a weir, sometimes called a *gull*, often flowed quite rapidly and was difficult for boats to pass over.

Paddle-and-rymer weirs held back smaller amounts of water than modern pound locks – commonly 1–2ft. In times of drought, opening was restricted to only twice per week. In the 1820s a flash sequence began at Lechlade on a Sunday afternoon and a second one was on the following Wednesday afternoon. Each flash of water travelled downstream at about 1mph, and took two days to reach Folly Bridge, Oxford. At least twenty flash-weirs were constructed between Lechlade and Oxford, and boats travelling downriver took full advantage of flash openings. A typical journey from Lechlade to Eynsham took 24 hours. By the eighteenth century there were perhaps thirty flash-weirs and as many as fifteen remained until 1862. A refinement at some flash-weirs was a sloping side channel with rollers (a boat-slide) for use by smaller craft so that the weir could remain closed and maintain water levels. The rusting rollers of a boat-slide still exist beside Iffley Lock, south of Oxford.

Pound Locks

Passage through flash-weirs (and flash-locks) was rather hazardous downstream and arduous upstream. Also, they were considered very wasteful of water. This brought about the introduction of modern pound locks, which were frequently sited adjacent to established weirs, as it was practical to employ the same keeper. Pound locks were named from the reservoir (or pound) of water created within a lock. The invention of pound locks is attributed to the Chinese in the tenth century, but those had rising gates. Only in the late fifteenth century did the familiar mitre-gate locks proposed by Leonardo da Vinci emerge. It is thought that the Exeter Canal was the first in Britain to use pound locks in the sixteenth century.

In 1624 Swift Ditch at Abingdon, only 1½ miles long, was reopened by the Oxford–Burcot Commission (1605) and they built a pound lock at its head. James I appointed, by Acts of Parliament, Barge Commissioners to improve the Thames, which had become almost impassable between Burcot and Oxford, and many goods had to be transferred to road transport. The king commissioned at least three locks in the 1620s to ease the transport of coal up-river after a sea journey. In 1790, today's channel at Abingdon was re-established with a pound lock; Swift Ditch was abandoned to nature and derelict by 1811.

The Oxford–Burcot Commission sponsored new locks, and the 1751 Thames Navigation Act empowered the Commissioners to control the non-tidal river, finance improvements, replace all flash-weirs with standardised pound locks, and construct towpaths. By the 1760s river trade was flourishing. A 1770 Act defined the dimensions of new (timber-built) locks, and set the tolls. Land purchase was made possible, and many lock-keepers were provided with cottages, though not until the end of the century did the Commissioners compel them to live in them. By a 1771 Act, twenty-two pound locks were commissioned and on the Upper Thames six were built by 1791.

Stone- or brick-built locks came into being in the 1780s, probably anticipating the increase of traffic with the opening of the Thames & Severn (T&S) Canal. Bridges at Godstow, Radcot and St John's, Lechlade, were earmarked for alteration to increase the headroom for larger boats. However, up until about 1820 most locks were built in places that suited mills and fish-weirs, many of which were still impeding navigation. Pound locks improved navigation, but the towpath was often not continuous. Some riparian owners refused to allow a towpath, so it had to cross the river, often by ferry or ford. But compulsory land purchases were made.

River Trade and Transport

A 1695 Act identified continual conflicts between navigation and milling, and in the mid-1700s the Upper Thames was said to have many shallow places. River levels constantly fluctuated, boats were liable to run aground, and even rowing boats could become stranded in summer. Transit between Oxford and London could take almost two months. When river trade was expanding and larger boats being used, the Gloucestershire historian, Samuel Rudder, wrote scathingly in 1779 that the Thames could not be relied on for the conveyance of merchandise. There were numerous plans to by-pass the Upper Thames; at first with canals, though speedier rail transport eventually achieved this. Nevertheless, the versatile railways became a boon to scores of Londoners who took excursions to Oxford for angling trips, towpath cycling (once pneumatic tyres had arrived) or to enjoy the pleasant pastime of boating.

After the T&S Canal opened in 1789 there were complaints that too little was done to keep open the Thames upstream of Lechlade; a certain amount of neglect led to increasing numbers of islands, shoals, and trees lying in the river. By 1811 boats of only 10-ton could no longer reach Cricklade, and in the 1860s even flat bottom pleasure craft could not travel far beyond Lechlade. Some dredging was done, particularly of fords, but funds to cleanse the river were not generally available.

Navigation along the Thames was paid for in tolls at weirs, locks, bridges and on towpaths, typically ½d (0.25p) per ton per mile. The system for weighing boats was a compound lever mechanism using the same principle as weighbridges for road vehicles, invented in the 1730s by John Wyatt of Birmingham. Tolls raised valuable capital for maintenance, but also raised transport costs significantly. In 1795 old locks were bought up, and new ones built. In the 1790s freight charges between Lechlade and London amounted to £1 5s (£1.25), but only £1 for the return journey. The Thames Commissioners abolished many tolls and in the 1860s the toll for barges was set at 2d per ton at each lock.

The Thames Conservancy came into being in 1857 and took over the duties of the almost bankrupt Commissioners who had managed the Thames since 1770. One of the first actions of the new board was to limit travel to 5mph downriver

Wharf-side merchandise.

and 4mph upstream. Canal barges were making regular week-long journeys between Brimscombe Port (on the T&S Canal) and London, but travelled only by day. There had been a law against working on Sundays but this was repealed in the Waterman's Act of 1827.

The Conservancy Board regulated river flow and made certain that boats of 4ft draught could reach Lechlade and the T&S Canal. They also made annual inspections of locks. Nevertheless, the inspectors constantly identified shallows where even a rowing boat of only 12in draught touched the riverbed. The Conservancy Act of 1894 was intended to make the river fully navigable and it was decided that all remaining flash-weirs should be dismantled. During the 1890s the Thames Conservators decided to build pound locks to improve navigation between the T&S Canal and Oxford; but the flash-weirs at Eynsham, King's, near Wolvercote, and Medley, near Oxford, remained in use until 1927.

Regardless of the problems, scores of trading vessels used the Upper Thames. In the early days, horse-drawn or sailing barges carried a variety of merchandise – grain, timber, hay, Cotswold stone, wine and cheeses, an easily portable form of milk in those days. By the early seventeenth century a typical vessel was somewhat like a punt but larger. They were open, flat-bottomed and square-ended, with a square-rig sail on a mast that could be lowered for passing beneath bridges. The towpath changed banks on various reaches. This relieved the strain on the horses' shoulders, and no doubt the horses welcomed cooling their feet in a ford during hot weather. Trace-horses, led by budding bargees, put their heads down against the heavy work, trudging through the countryside, or passing houses and the occasional wharf.

Nineteenth-century barges carried between 30- and 60-ton cargoes. Many were typically spritsail rigged and included a cabin. Between Lechlade and London the journey took about a week, but perhaps nearly two weeks to return. Horse-drawn barges were slowly superseded by diesel-engine craft. Even electric boats made an appearance on the waters of the Thames before the end of the nineteenth century. The first, called *Electricity*, was about 8m in length with a ton of batteries on-board; it could carry up to twelve passengers. It was innovative at the time, but no sooner than a century had passed the advantages of pollution-free boating had become accepted.

Canal Navigation

The Industrial Revolution brought rapidly increasing trade and new life to towns and villages, but many roads and ponderous waggons were still inadequate for economic transport of bulky goods. Then, about 250 years ago, canals were built with great zeal; George III's reign was exceptional for its 'Canal Mania'. Although canals were an improvement on the whims of rivers, they were not a total solution. Users of canals were soon to discover such problems as water freezing over in winter or banks drying out and leakage in hot summers.

One construction that began in the 1780s was to have a huge effect on Thames navigation. The 28-mile Thames & Severn Canal would provide cheap transport, despite tolls, for such goods as coal, timber and building materials such as Welsh slates. The former dependence on local materials – timber, stone and thatch – was completely altered by canals, and later rail transport. Visualise the impact of a new waterway cutting through the countryside, not only on the landscape but also on the people who lived nearby. The sinking of clay-pits, and building of brick kilns began for production of massive numbers of bricks for bridges and locks; all done with manual labour.

The T&S Canal was an important west–east route from the Severn to the Thames, and it helped to reduce some tolls. It also opened a link between the Midlands and London; the latter being seen as the largest potential market in England but physically separated from industrial centres of the Midlands and the north-west. The canal linking Brimscombe Port on the Frome to the Thames was supplied mainly with water from Thames tributaries. The Canal Act authorised taking water from rivers with compensation for riparian owners who were inconvenienced by the loss of current in the streams. Claims came typically from owners of water meadows, and millers who wanted to ensure a regular supply for milling flour, which keeps less well than grain. Minimum mill activity was usually between spring and June when river flow was slowest. For example, in late summer it was common for the Churn to have 20 per cent of full flow.

The canal route through the Cotswolds included a 2¼-mile tunnel at Sapperton and in 1788 George III visited the construction site. East of the tunnel the fall of the canal was gentle compared with the steep climb from the Frome valley. In the first 6 miles eastwards no locks were needed, and only sixteen locks altogether between the summit and Lechlade for a total descent of about 130ft (42m).

Six and a half years elapsed before completion of the canal, with the first boat going through Sapperton Tunnel in April of 1789, though the grand opening of the whole canal was not until November that year, when there was a bonfire party at Lechlade. There were many benefits from the canal; for example, coal deliveries (almost entirely from the Forest of Dean) at Cirencester (on a short canal arm from Siddington) were reduced by 6s (30p) per ton.

Canals were the best way of transporting bulky goods well into the 1870s. The Thames Commissioners responded to increased demands put on the river by completing six pound locks before 1791, and removing some older flash-weirs. They built numerous timber footbridges, doubtless where fords were grubbed

The imposing eastern portal of Sappeton tunnel (SO966006) was restored by the Stroudwater–Thames & Servern Canal Trust, and the plaque above the arch was unveiled by the Rt. Hon. Earl of Bathhurst in 1977.

out. The journey from Brimscombe Port to London took under two weeks; but by the late eighteenth century fast mail-coaches between Bristol and London completed the journey in about 24 hours. The declining use of the Thames and the T&S Canal was partly due to a burgeoning of a more-widespread canal system and the growth of railway networks from the 1840s, and by 1862 there was very little traffic on the T&S.

The Oxford Canal opened in 1790 with a 42-mile shorter route between the Midlands and London, halving the cost of transport of coal into Oxford. The Kennet & Avon Canal opened in 1799 and linked Bath on the Avon Navigation with Reading on the Thames. It provided easier communication between west and east coasts, severely threatening much of the T&S Canal trade originating in Bristol and the south-west. Ironically, canals began transporting the materials for building the railways, and by the beginning of the twentieth century the railway link between Bristol and London reduced travelling time to about three hours.

Today, much of the T&S Canal is derelict or completely wild. However, in the last quarter of the twentieth century, a feasibility study was conducted for restoration of the canal, and an energetic, ever-swelling band of enthusiasts is reviving the historic waterway. Already a number of short sections and the portals of Sapperton tunnel near Cirencester have been restored.

three
The Future

The Upper Thames gives great pleasure to all. It is the haunt of mallard and moorhen, heron and swan. The shy otter is being seen more frequently in the upper reaches after the years of fishermen's persecution caused it to desert the area around a century ago. Whether it stays depends on many ecological factors, but the otter is probably no longer an endangered species. Sadly the water vole, Ratty in *Wind in the Willows*, is possibly on the brink of extinction, though riverbank repairs with living stems of withy may help to conserve its habitat.

The river attracts scores of boating enthusiasts, and keen oarsmen who emulate those striving along its course at the summer regattas of Oxford and Henley. Just occasionally, an otherwise picturesque stretch of water is marred by numerous moored cabin cruisers, some seemingly left to rot and create rainbow trails of oil across the water's surface.

Centuries ago the Thames and its tributaries teemed with fish, and riverbank inns were frequently named after the plentiful trout, perch and pike. Over the centuries the lower reaches of the Thames became so polluted that it was deemed lifeless; but now it is much cleaner and many wildfowl and fish have returned. The tributaries Churn, Coln, Leach, Windrush and Evenlode rise in the limestone hills and are ideally suited to the trout that visit the upper reaches to spawn. Fishing is popular despite the loss of salmon about 150 years ago, though in the last twenty years, with cleaner lower reaches, the somewhat choosy fish with an extraordinary life cycle has been tempted

Thames-side fisherman.

Walking riverside footpaths or travelling by dinghy can reveal much history of the Upper Thames. The gentle pace makes for a deeper appreciation of bygone days, enhanced by the delights of today's countryside.

Numerous new footbridges were built for the Thames path.

to return, and one was seen near Oxford in the late twentieth century. It is to be hoped that the mysterious method by which salmon find their way to the spawning grounds will mean larger numbers return to the silt-free Upper Thames shallows as time goes on.

section one

From Source to Cricklade

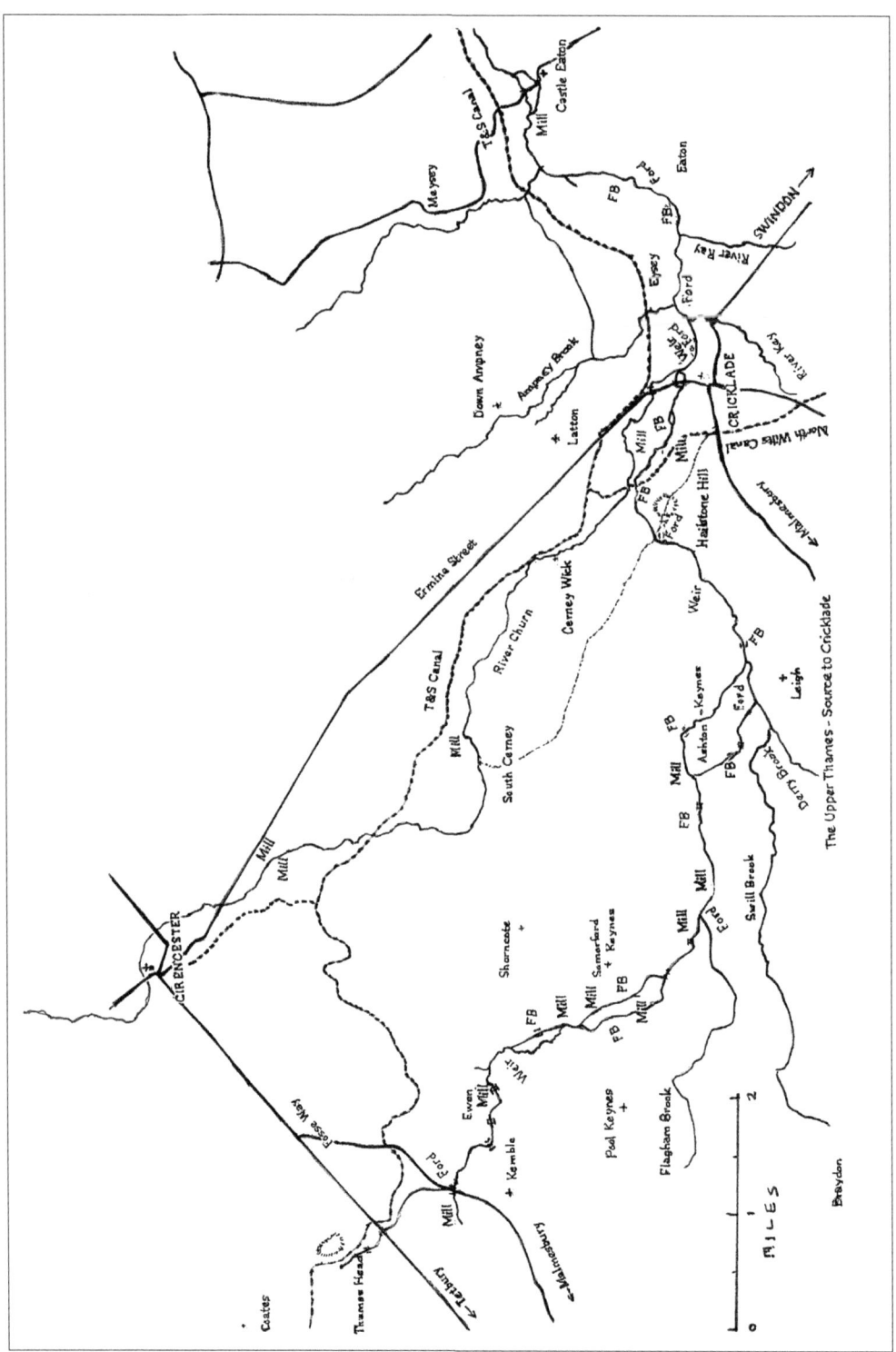

The Upper Thames – from source to Cricklade (12 miles).

The Seven Springs, the reputed Thames source since at least the fourteenth century.

Thames Head to Cricklade

The Thames rises in the limestone Cotswold Hills of Gloucestershire where scattered villages sit amid a landscape of arable fields and pastures with countless sheep producing highly prized wool. Its name is ancient, first mentioned by Caesar in 51BC as Thamesis. The name stems from Sanskrit, an Indo-European language, probably used by the first inhabitants, and means 'river that swells'. It is also called the Isis, a practice that began in the fourteenth century around Oxford, but this also derives from Thamesis.

The river has no mountain torrents because the source is only 370ft (110m) above sea level. In a shallow, wide valley south of Cirencester the spring (ST985987) rises at the foot of Trewsbury Castle, an Iron Age fort. Thames Head is marked by an inscribed granite block erected by the Thames Conservators in 1974. A few paces in front of the monument is a shallow dell of damp stones and a sentinel ash-tree stands with oak and field maples nearby. A reclining figure of Neptune as bearded 'Father' Thames once stood here but was moved to St John's Lock, Lechlade. He began his public life at the Crystal Palace Exhibition in London where Raffaelle Monti had his sculpture displayed.

The spring is invariably dry causing belief that the true Thames source is at Seven Springs, north of Cirencester. This source becomes the River Churn, a Thames tributary that never fails despite rising at a higher elevation. It is also further away from their confluence, which in geographic terms gives it superiority. Clearly the Churn is the stronger stream rising further from the sea.

The Gough Map of Great Britain, brought to light in 1774 but dating from around 1350 and now in the Bodleian Library, shows the Thames at Seven Springs. In 1760, Seven Springs was called Thames Head and an early twentieth-century postcard fostered this belief, endorsed by an inscription at the spring: *Hic Tuus O Thamesine Pater Septemciminus Fons*. In the late 1430s the waters of the Thames were said to flow at Cirencester; however, in the sixteenth century John Leland said that the source was near Kemble.

The 'official' source near Trewsbury Castle may have resulted from Earl Bathurst's dream from the 1700s of integrating a proposed canal and the Thames within the landscape of his park at Cirencester. The canal was an extremely important development for the Thames as a highway, causing many changes to ancient crossing places along the river.

> *Coates parish is the official location of the source of the Thames. The earliest document referring to Coates is dated 1175, when Cotes was usually taken to mean 'a shelter' – as for sheep. By the thirteenth century it had a manor estate held jointly by Ralph Paganel and Elye Cokerel. The village church of St Matthew, of Norman origin most evident in the south doorway, has a Perpendicular tower dating from before 1361.*

Moving east the river's course seems blocked by the embankment of the Fosse Way, but there is a culvert to convey any water (ST985990) making this the first crossing place over the Thames and the oldest datable crossing. The Fosse Way was built by the Romans to ease distribution of military reinforcements along a route between Lincoln and Exeter. The road's significance is identified in Anglo-Saxon charters and an entry in the *Anglo-Saxon Chronicle* for the year 880 tells how a Great Danish army moved from Chippenham to Cirencester. Although their route is unknown it is possible they crossed the Thames on the Fosse Way.

Meadows around the Infant Thames, north of Kemble.

> *A short distance along the Fosse Way embankment and almost lost in undergrowth, but protected by a retaining wall, is an ancient boundary stone. The Hoar Stone, possibly the oldest piece of masonry along the river valley, is apparently of Saxon origin. It was referred to in a land grant at the time of King Athelstan and served as a boundary marker for Malmesbury Abbey lands. It has been cut on one side, maybe to form a mounting block, and it is now well clad in shaggy moss.*

Beyond the Fosse Way culvert is a patch of damp ground. The stream meanders through the meadows where a strong spring emerges. This is Lyd Well, which was utilised by the builders of the Thames & Severn Canal. According to Geoffrey of Monmouth in his twelfth-century *History of the Kings of Britain*, Lyd became king of the Britons and renamed Trinovantum as Caerlud, later called London. Lyd's burial place is said to be in the capital at Ludgate and it is fitting that the great river links Lyd Well and London.

Following a footpath towards Kemble, the course of the T&S Canal is on our left. In the early part of the twentieth century a ford (ST988986) over the Thames gave access by track beyond the canal to farmland. Only 150 yards south of the ford there was a wind pump that raised water for the canal.

The Thames & Severn Canal would have a major impact on the use of the Thames and its crossings. It was a major feat of engineering, conceived in the 1660s though not officially surveyed until 1781. The T&S would be a link between the Stroudwater Navigation of the River Frome, a tributary of the River Severn, and the Upper Thames at Lechlade. Robert Whitworth, who had expertise as an engineer and draughtsman, as well as being a protégé of the canal engineer James Brindley, was a champion of canals and was asked to conduct a preliminary survey of the Upper Thames. Promoters of the T&S, many in the West Midlands, firmly believed in inland navigation and there were at least sixty-three proprietors/shareholders.

Royal Assent for the canal was given in 1783 and a line was decided, necessitating a tunnel almost 2¼ miles long to be cut through the Cotswolds. The tunnel was at that time the largest engineering project in the world and the overall cost of the canal was nearly a quarter of a million pounds. The Sapperton tunnel took four years to construct; navvies used mining techniques, hand-tools and gunpowder, with their only light being from candles. Some lost their lives in accidents and today we should hail them as heroes of the construction. A little south of Coates is a restored section of the canal. The renovated eastern entrance of the tunnel, located about 200 yards (180m) off the Coates to Tarlton road, was built 15ft (5m) wide and high, with columns, roundels and a Cotswold-stone façade with niches probably intended for figures of 'Father' Thames and the Lady Sabrina, goddess of the Severn.

Canal maintenance was carried out by watchmen who were paid 9s (45p) per week. Most watchmen had cottages, but between Chalford and Lechlade there were five three-storey roundhouses, which allowed a watchman an excellent view of the canal and any approaching boats. The roundhouses had stabling for barge trace-horses and watchman's accommodation upstairs. Three of the roundhouses had a specially constructed roof in the form of an inverted cone lined with lead to collect rainwater for domestic use, which supplemented the meagre natural supplies.

The T&S opened in 1789 and served traders on the Thames for about 130 years. The last barge passed through Sapperton Tunnel in 1910. Traffic between Bristol and London had steadily diminished. An all-time low for toll revenues occurred in 1816 when the canal was in a poor condition and, despite many improvements, by 1839 barges were almost non-existent on the upper reaches of the Thames where boats frequently ran aground. By the 1860s the Upper Thames reaches were virtually un-navigable. For some time movement of goods had been predominantly eastwards, which caused logistical problems and many westbound boats were often only partially filled. This was also a serious period of agricultural revolt, low wages and near starvation.

A two-arch bridge for the A429 (ST991979), formerly the site of Clay Ford, near Clayfurlong Farm, Kemble. The road links the ancient towns of Cirencester and Malmesbury.

> *An 1882 Act of Parliament applied for the canal to be closed but was opposed. Competition from the railways was steadily growing, but there was much discontent from regular traders on the Canal and the Thames about lack of maintenance, as they argued that water transport was cheap and convenient. Initially the railways seemed advantageous but frequently, due to many places not yet being served by rail, journeys took longer and were more costly for bulky goods. The final closure notice came in December of 1893; there was a short reprieve but by the 1920s there was practically no traffic at all. In 1924 authorisation was given for the 19-mile section east of the tunnel to be abandoned. The towpath soon became virtually a right-of-way, except where Nature established tangled brambles and shrubby trees. Untended stretches soon filled with evil-smelling sludge, though parts became havens for wildlife; a sad end to an historic partnership between a man-made watercourse and the Thames.*

The infant Thames flows in clear shallows through the north of Kemble parish where a sheepwash was located. It passes beneath a bridge constructed in about 1791 at Clay Ford (ST991979) near Clayfurlong Farm on an ancient route linking Cirencester with Malmesbury. Malmesbury Abbey under Abbot Aldhelm, a member of Wessex royalty, owned farmland around Kemble in 675. The

Cattle cooling their heels in the Thames near Clayfurlong Farm; these shallows are very close to a former sheepwash.

Domesday Survey identified two mills at Kemble worth 15s. A mill on the 1828 OS 1in map was plausibly on one of the sites, though a contemporary painting shows a windmill at a similar location. The watermill would have been within a mile of the source, and possibly only worked during winter months, as the river here often runs dry in summer. Until 1897, when the Gloucestershire boundary was moved, Kemble was in Wiltshire.

At Clay Ford there was also a three-arch stone bridge for the Cirencester Branch of the Cheltenham & Great Western Union Railway. The branch line was single-track broad gauge when opened in 1841 but converted to standard gauge in 1872. Local newspapers of the day advertised cheap railway excursions: especially to the Crystal Palace Great Exhibition where coincidentally Raffaelle Monti had his sculpture of 'Father' Thames. The line was popular with passengers and was used for transporting coal, building materials and agricultural products until its closure in 1965, which caused some bad feeling in Cirencester.

Railways came early to this area and the flamboyant Isambard Kingdom Brunel was the architect for Kemble station, which was not opened until 1882 despite the Cirencester branch coming into operation forty years earlier. The railway crossed land belonging to the owners of Kemble House who initially refused a public station on the estate and furthermore stipulated that the railway should not be visible from the House, necessitating a 400-metre tunnel. In due course a station of sorts came into use, but only for interchange of passengers to and from Cirencester. Later, Brunel's station was opened for passengers from the village. Today its structure is largely unchanged.

Kemble was Kemele in 682 and archaeologists have located in the neighbourhood Saxon burials dating from the sixth century. The origins of Kemble's name are uncertain; it may derive from a Celtic god or possibly from an Old British (Celtic) word meaning 'boundary'. Place names in Old British surviving through the Roman era are rare, though Asser, King Alfred's biographer, wrote that the 'old language' was still spoken in the ninth century. This area, inhabited from prehistoric times, is said to be on the southern boundary of the Iron Age Dobunni tribe, though it also abuts the Fosse Way, which may have served as a notional western frontier in the early Roman era.

After the Romans' departure, which was no doubt protracted, the mixed race that remained was led against invading Anglo-Saxons by characters like 'king' Arthur, with occasional victories. The immigrants asserted their power across southern England during the sixth century, and the Battle of Dyrham (north of Bath) in 577 drove a wedge between the Britons beyond the Severn and those in the south-western peninsula. The old Roman towns of Bath, Gloucester and Cirencester were captured, and Mercian rule came to the Cotswolds. Saxon occupation of the Upper Thames valley is known from their fifth-century cemeteries. The Gewissae ruled the Upper Thames region from about 600 with a major 'royal' settlement near Benson, and although two Gewissae kings had Celtic names archaeologists believe they were Anglians. In the seventh century, Mercians captured the Briton's stronghold of Dorchester-on-Thames, sparking numerous disputes along the Thames 'frontier' with the West Saxons.

The Thames, now at least 1ft deep, meanders through farmland and under Parker's Bridge (ST996974). Many years ago this bridge included a long causeway over the swampy land. Before reaching Ewen, the river passes Mill Farm (ST999973), the site of the second mill on this young stream. Adjacent to Home Farm is a modern bridge (SU004973) but the water is shallow over a gravelly bed and there was doubtless once a ford. In 931 Ewen was *Awelme*, meaning 'plentiful spring', and a local spring joins the river near a modern weir (SU007973). Isaac Taylor's map of Gloucestershire identifies the settlement as Yeoing.

The river passes field and copse and flows into open country; it divides (SU010971) where a leat was constructed long ago for another mill – also sadly

The mill leat outfall near Upper Mill Farm, the site of Washburn's Mill in 1830 (SU013963) and a sixteenth-century weir. The small timber footbridge crosses the lower end of the mill leat and a little further downstream of the mill-pool the leat rejoins the 'old' river.

gone. At least nine mills worked between Kemble and Cricklade, indicating the river's value even in these upper reaches. During the nineteenth century steam-driven mills brought a demise of watermills. Both were eclipsed by twentieth-century use of electricity.

The old course of the river was usually left untended by the weed-cutters, but the fast-flowing mill-leat runs beside a meadow, which shows possible remnants of drainage channels between the higher leat and the old river. These could have flooded the meadow to promote early summer grass. The leat passes under a farm track bridge (SU012965) built when land-holdings were divided by the new leat, which arrives at Upper Mill Farm where the water gushes through an old millrace.

This was Washburn's Mill in the nineteenth century and noted for a remarkably large millwheel. In the sixteenth century a weir was recorded here, maybe when the mill was built. A small timber footbridge (SU013963) crosses the lower end of the mill-leat and a little beyond the mill pool the leat rejoins the old river, but the river divides again for a small channel to Somerford Keynes; this is crossed by field paths of uncertain history. A plank footbridge (SU015954) bends beneath the tread, and another one is at sited at SU019951.

Following the modern watercourse (leat) through pleasant country we arrive at Old Mill Farm (SU012957), which was Pool Mill in 1327. All that remains now is a reed-filled mill-pool below a small stone-built duct off the swiftly flowing leat. In 1327 John atte Mulle resided at Mill Farm. Washburn's and Pool mills are both shown on the 1828 OS 1in map.

> *Poole Keynes village is some distance to the west; it was* Pole *early in the tenth century, only acquiring its suffix when, in about 1327, a daughter of John Maltravers, lord of the manor, married Sir John Keynes.*

A small stone bridge at Old Mill Farm – former site of Pool Mill 1327–1830 (SU012957). John atte Mulle resided at Mill Farm in 1327.

The river passes Kemble Mill (SU013952), dating from before the sixteenth century, which is now a private residence. The parallel water courses reunite at Somerford and pass beneath a stone bridge (SU019948), doubtless where a ford was once located. Somerford's name, first noted in 683, suggests it had a 'summerford'. During a late nineteenth-century drought this stretch of river was said to be less than 3in deep, making it easily fordable. The Domesday Survey records a mill at Somerford.

This area is called Neigh Bridge; a name recorded in 1327 but altered to Neybridge for a while after 1591. It is now almost surrounded by lakes in former gravel pits – the water seeping in from rivulets in the neighbourhood. Extraction of valuable sand and gravel began in the 1920s and continues today. However, gravel extraction has had a detrimental effect on recovering archaeological information

Somerford Bridge (SU019946), Somerford Keynes, first noted in AD 683. Somerford doubtless had a 'summer-ford'.

hidden beneath the fields but known to exist from aerial photography. In fact, this part of the Cotswolds is one of Britain's richest archaeological areas. There have been some interesting finds, such as a mammoth skull complete with teeth, and thought to be over 50,000 years old. The gravel-pit lakes extend over some 30-odd square miles known as the Cotswold Water Park, which provides a wide range of recreational facilities within the largest area of man-made lakes in Britain. However, large areas of open water allow evaporation to affect groundwater levels.

> *Documents show that the Saxons established Somerford manor. In about 685 Berhtwald, a nephew of Ethelred, king of Mercia, gave about forty hides of Somerford land (one hide of land could support one family) to Aldhelm, abbot of Malmesbury. Aldhelm, who was subsequently canonised, was an energetic missionary who wrote on the error of the Celtic Church in calculating Easter Day. He is said to have been a talented harpist and composer of sacred songs.*
>
> *In 1211 the feudal landlord was William de Kahaines and in 1291 Somerford was appended Keynes (from the family who originated in Cahagnes, Normandy). All Saints church has a small Saxon-style north doorway, some eleventh-century sculpture, a fifteenth-century rood screen and an eighteenth-century Gothic tower. There is also a monument to Robert Straung (1645) in flamboyant dress and wig, leaning on one elbow as if reclining in the grass. Somerford Keynes was in Wiltshire until county boundary changes in 1897.*
>
> *In a rural setting to the north-east is the tiny hamlet of Shorncote, mentioned in Domesday Book as Schernecote meaning 'a mucky place', which was transferred into Gloucestershire in 1897. East of the village, near a track to South Cerney, archaeologists found a Bronze Age burial ground – further proof that the area has been inhabited for at least three millennia. All Saints church dates from about 1170 and, despite the isolated setting, hints of a former importance. The church has a double bell-turret with a bell made in 1706. The interior has carved bosses for the chancel's waggon roof and murals above the chancel arch, which itself bears Royal Arms cut into the stonework.*

The Thames has achieved a significant depth and width where it passes beneath a road bridge (SU018947). Nearby, Mill Lane (bisected by a newer road) leads to the former Somerford Mill (SU023944) on Lower Mill Farm Estate, and then on to Somerford Common. Mill Farm, first noted in the 1320s, had a fish-weir (*fysshwear*) in the sixteenth century and only ceased milling cattle feed in the

1960s. Due to gravel extraction, the Thames's historic course is ill-defined from here as far as Waterhay Bridge.

The footpath between water-filled gravel-pits leads to Ashton Keynes, crossing a footbridge (SU026943) where there was a gravelly ford until quite recently. A little further on, the river passes beneath a bridge (SU028940) at the former site of Skilling's Mill (1828 OS 1in map). Flagham Brook (first noted in 1633) once joined the Thames nearby at SU036942, but the stream is now lost amid the flooded gravel-pits.

The Thames now enters Wiltshire, a land-locked county. The early inhabitants, who were overcome by the Romans, have left a few indicators of their language in the names for the forests and rivers, but the county is most noted for its outstanding prehistoric sites such as Stonehenge and Avebury. As our path enters Wiltshire we cross a boundary thought by archaeologists to be over 3,000 years old. Excavations (reported in British Archaeology) *prior to gravel extraction revealed a series of pits, which closely follows the line of the present boundary. New evidence has shown that the boundary pre-dates the Saxon era and also that the Romans acknowledged it when they laid out field and property boundaries. This discovery is by no means unique in England.*

Saxon colonisation began in the late fifth century and by the early seventh the West Saxons had forged a kingdom (mostly Christian) around the Upper Thames. They began a network of parishes, probably based on earlier Iron Age territories. Medieval wealth came from lucrative farming largely run by religious communities who reared thousands of sheep, initially producing milk and later valuable wool. Woollens manufacturing was carried out in numerous weaving mills. Other small industries grew in north Wiltshire; for example, from the sixteenth century gloves were made by women and girls in the area, working long and hard in their cottages for a few pence per week. The leather, mostly from sheepskins, was produced in tan-yards using tannin made from locally stripped tree-bark. By the nineteenth century the trade extended to making gaiters and leggings.

Wiltshire is still predominantly rural with cattle farming and some pig rearing, though much less so than in earlier times. A significant industry in the past was ale brewing and this continues. There is a scatter of settlements and small towns whose economies have long relied on country markets; only Salisbury and Swindon seem of any great size, though a few towns near the M4-motorway are rapidly developing.

Today, security from wild animals has generated interest in wildlife preservation and bird-protection. Despite improved food production, many of us have developed a taste for 'weeds' rekindled by a longing for the countryside.

Site of Skilling's Mill in 1828 (SU028940), near Ashton Keynes.

Before arriving at Ashton Keynes the Thames flows on an eastbound course between the many lakes of Keynes Water Park, which provides a variety of attractive habitats for wildlife. The Thames enters the village near the church, passing beneath a low bridge (SU041942) over the stony riverbed. It then comes to Ashton corn mill (SU044942), identified in Domesday Survey but not used since 1910. Today the mill-house is a dwelling with millstones set into the walls of the south and north gables. The Domesday entry listed a mill worth 5s (25p). Beyond the mill the stream flows picturesquely between clipped lawns and then comes to the ancient stone-built Gumstool Bridge (SU045942) named in the days of ducking stools for miscreants.

The river's passage through the village is rather complex. It divides and a shallow, often dry, channel flows eastwards beneath a bridge (SU045943) bearing an illegible stone-cut plaque. Two modern housing areas (Richmond Court and The Leaze) have bridges over this channel, which turns south near a track to Kent End Farm that was recorded in 1327. The footpath continues to Waterhay Bridge, but a footbridge (SU052942) of massive stone slabs takes another path back towards the village. In this part of Wiltshire, large stone slabs are often placed

Gumstool Bridge, Ashton Keynes (SU045942); the unusual name is synonymous with the ducking-stool once used for transgressors of the law.

on end to form boundary walls. A bridge (SU053942) takes the road to Rixon Farm (first recorded in 1603). The path to Waterhay Bridge crosses footbridges at SU054938 and at SU056937.

Also from Gumstool Bridge a stronger stream, in places called Gosditch (presumably once Goose-ditch), flows southwards past the site of a former horse-washpool and beneath a variety of little bridges providing access to private properties. It flows on towards Happy Land and reaches Oaklade iron-railing bridge (SU049934), the 'lade' no doubt because there was once a ford, where later a three-arch stone bridge stood dating from the mid-1820s.

> Ashton Keynes, first mentioned in 880, was described in Domesday Book as 'a settlement with ash-trees'. In 889 King Alfred gave land here to his daughter, Aethelgifu who was abbess at Shaftesbury. Henry Kaignel held Ashton Manor in 1242 and William de Kaynes held it from 1256. There are three medieval preaching crosses in the village, and a fourth in the churchyard serves as a First World War memorial. West of the village, set on raised ground, is the old moated-manor site, where historians say monastic buildings were enclosed within a moat. Holy Cross church is adjacent, and the Holy Cross is above the south porch dating from 1490. In the nave there are carved oak tie-beams resting on sculptured stone corbels and a tub font decorated in an early style with chevrons and foliage.

A little south of the village the Thames is joined by Swill Brook and Derry Brook (Grenebourne in around 1368 or Sandburn around 1651). High Bridge crosses the Swill Brook and nearby a track crosses the Thames at a ford (SU054934), though a small timber footbridge now allows a dry-shod crossing on a route that may once have had greater use. In the meadows beyond the ford, crop-marks, which form linear features, have been identified – perhaps remnants of ancient trackways or silted-up drainage channels? Drainage Commissioners applied the Thames Valley Drainage Acts in the 1860s and 1890s, and they taxed holders of low-lying land. Failure to pay the tax meant withdrawal of voting rights for the local Board.

The Swill Brook was known in the mid-tenth century as *Bradenbroke* – a name in use today at Braydon Brook Farm (near its source) in an area once noted for Braydon Forest medieval pottery. Braydon Forest covered a vast area, which until the mid-thirteenth century extended northwards to the Thames and east to the River Ray. By 1330 its northern boundary had shrunk back to south-west of

Ashton Keynes High Street runs alongside the Thames, which flows beneath numerous little bridges giving access to residential properties. (Courtesy of www.old-england.com)

Cricklade, except along the Thames where it still reached Hailstone Hill and the land of William atte Brugge.

The strengthened Thames heads for Cricklade passing under Waterhay Bridge (SU060933) with painted iron railings. The present bridge may date from the late 1890s but there was previously a masonry structure. Navigation for barges is believed to have existed here, but mills and their weirs and bridges downstream would have hampered most boats. The name Waterhay, of Anglo-Saxon origin, means 'woodland farm'. Any village is now lost but farms survive at Upper Waterhay.

A short distance downstream near Brook Farm, which dates from about 1680, there is a relatively new iron-girder and planks footbridge (SU065933). The farmland is largely unimproved which has allowed snakes-head fritillary plants to continue to thrive.

The murky river turns north-east and on its northern flank are water-filled sand and gravel pits, home to many wild birds. A farm track crosses the river via Bournelake Bridge of two arches (SU074939) – one of stone and the other of brick. This is very close to the former Bournelake Weir (1906), and in the 1300s Henry atte Bourne had land here. Bournelake Farm was first recorded in 1630.

A bridge near the former Bournelake Weir of 1906 (SU074939); in 1333 Henry atte Bourne had land nearby.

Soon the Thames reaches a crossing near Hailstone Hill. Saxton's 1576 county map shows Hawston, sounding rather like Hailstone, which is said to derive from Halagheston used in 1228. Old English words *halig* and *hlaw* refer to a hallowed mound. Approaching the crossing, the wide valley plains are revealed as morning mist evaporates in the sun. A timber footbridge (SU079945) soon comes into view; at one time there was a ford, though a bridge is mentioned as early as 1140. This crossing has long been a route between South Cerney and Cricklade via Stones Lane. By 1779 there was a stone bridge. The path northwards enters Bridge Ham and then Thames Furlong, a relic of medieval ploughland, which pre-dates the 1814 Enclosure map. Continuity of use here confirms a very ancient route.

In a short distance the river, weaving a course through reeds, comes to a dilapidated footbridge of planks (SU083947) above some decaying supports, which looks distinctly unsafe, but might perpetuate a crossing formerly provided by a weir. A little further on is a stronger bridge (SU083947) built in 1883 for the Midland & South-West Junction Railway. The stone and brick abutments remain but are now linked by a gravelled bridleway supported on steel girders. The

The footbridge near Hailstone Hill (SU079945) at one time was the site of a ford, although a bridge is mentioned as early as 1140. By 1779 there was a stone bridge. This crossing is on a route between South Cerney and Cricklade via Stones Lane.

railway served Cirencester (Watermoor Station) and Cricklade until 1964. In the disturbed days of the Second World War this bridge carried trainloads of troops, wounded soldiers, and munitions from the Midlands to Southampton.

> *During the first quarter of the twentieth century Watermoor Station (now a car park) was the main despatch point for large quantities of milk from surrounding farms conveyed in the once familiar two-hundredweight churns. This was also a gateway to the Cotswolds competing with the GWR line through Kemble. Steam trains travelled northwards and passengers, glancing from the train, might have thought their train in flight above the young river.*
>
> *South Cerney has Saxon origins. Its name relates to the River Churn, a Thames tributary first mentioned in around 800 as the Cyrnea. Another place name possibly derived from the Churn is Cirencester, whose Roman name was Corinium, probably formed from its Old British name. According to archaeological evidence this area was inhabited in prehistoric and later times. Finds include late Ice Age axeheads and a thirteenth-century penny coin of Alexander III, founder of the Scottish nation, was unearthed when a pipeline was laid in the 1980s.*
>
> *Before the Norman Conquest, Cerney manor belonged to Bishop Stigand who sits at Harold's left in the Bayeux Tapestry. He was Archbishop of Canterbury from 1052 until deposed in 1070. William of Normandy wanted Abingdon Abbey to hold Cerney manor, but it became the estate of Walter, son of Roger, Sheriff of Gloucestershire.*
>
> *South Cerney has attractive bridges over the Churn, a street called Bow-Wow flanked by the river and an old mill leat. All Hallows church is mostly Norman; the south doorway arch has flamboyant beak-head decoration and other carved stonework, and inside there is a twelfth-century wooden rood fragment depicting Christ crucified. Cerney had an airfield before the Second World War and the RAF occupied it until 1971. Today it is a gliding centre and army depot.*
>
> *To the south-east is Cerney Wick; possibly where a wic (or dairy farm) was worked on land granted in 1220 to Nicholas de Valers whose monument is in Down Ampney church.*

The Thames' course is unclear until joined by the River Churn at Cricklade. Daniel Defoe commented that the Thames was four streams at Cricklade. The Thames and Churn were used as feeders for the T&S Canal, which followed a course near to the Roman Ermine Street. The Churn skirts the north-eastern side of North Meadow, adding its significant flow to the Thames shortly after passing beneath Weavers Bridge (SU103940), rebuilt in 1929. Beside this bridge on the ancient Ermine Street is Turnpike Cottage where a pulley block, for lifting goods to its loft, projects near its eaves.

> Between the two rivers lies the vast North Meadow, especially renowned for an outstanding springtime display of the purple and cream flowers of snakes-head fritillary, which only colonises ancient unploughed meadows. The common pasture, used since at least the fifteenth century, is grazed in the traditional Lammas method and the meadow is claimed to have three-quarters of this country's fritillary plants.

In 1810 a scheme was afoot to build the Severn Junction Canal linking the T&S Canal near Ewen to the then recently completed Wilts & Berks (W&B) Canal. It would be 'a great public utility' and benefit movement of Forest of Dean coal, stone and many other commodities. The Upper Thames was receiving poor attention to improvements by the Thames Commissioners, and in all probability would be 'by-passed'. The Upper Thames would be abandoned in favour of the new canal.

The Thames Commissioners had been promising new locks upstream of Oxford for twenty years, but the new link canal could benefit the T&S by giving uninterrupted navigation to Abingdon. However, a strong opponent was Edward Lovenden, a Thames Commissioner and shareholder in the T&S Canal Company. He refused to be associated with the project. He owned Buscot Lock near Lechlade, the first river lock after the T&S Canal. After long discussions by the respective parties, a canal, renamed the North Wilts Canal in 1812, was planned between the T&S Canal Latton Lock and Swindon where it would join the W&B not far from its summit.

The North Wilts Canal officially opened in April 1819, but concern was expressed that operation of NW Canal locks would cause water to be drawn from the T&S, and double stop-gates were fitted at Latton in 1826 to prohibit the passage of boats when the NW Canal level was below that of the T&S. Surprisingly, the Upper Thames benefited from the NW Canal and began to experience a revival. The Thames Preservation Act of 1885 identified the river for pleasure craft – this was the advent of it becoming a popular recreational waterway.

> The NW Canal was beset with difficulties, not least a shortage of water, but it carried Forest of Dean coal and stone, Midlands salt to Wiltshire, and Somerset coal to Cricklade for the villages around. Its most prosperous decade was the 1840s. In 1841 the NW Canal was in danger of being put out of business when the railway came to Kemble. In the early twentieth century the W&B itself became un-navigable and was closed in 1906.
>
> Latton – Latone in Old English – had a medieval manor, possibly at Manor Farm to the north-east, but the main manor became Latton Court. At the village entrance is the stump of a medieval preaching-cross, restored in 1982. Aerial photographs show settlement traces in the fields to the south-east, suggesting habitation since prehistoric times.

At the northern end of North Meadow a footbridge (SU087948) over the Thames once carried a North Wilts Canal aqueduct and towpath. The brick buttresses remain and in the river are hints of the foundations for the former arches. The footpath continues to the former Latton Basin adjacent to which stands Wharf Farm, formerly one of the nine places east of Sapperton Tunnel where canal goods were discharged or loaded. In the T&S Canal heyday of the 1790s a canal agent

Footbridge (SU087948) where the North Wilts Canal formerly crossed the Thames near Cricklade.

had a house at Latton and drew an annual salary of about £50 but, due to falling wages in the late nineteenth century, the lock-keeper chose to supplement his meagre income by selling garden produce. In the late eighteenth century, wharf men received typically 1s 6d (7.5p) per day, but within ten years pay was raised to 11s (55p) per week. In 1871, a weir was recorded hereabouts but its location is now unknown.

South of North Meadow there is a modern weir gauging Thames flow. Adjacent to it are some run-down farm-buildings, behind which an unfriendly landholder often blocks off the bridleway back to the footbridge over the old canal route. A footpath bridge (SU095942) perpetuates a crossing near the site of the former West Mill, which existed by 1228, but was closed down by the Thames Conservators after working for about 700 years. In 1607 the mill was the highest navigation point – 44 miles from Folly Bridge, Oxford.

The Thames flows around the northern edge of Cricklade, passing beneath Town Bridge (SU102940). There is cartographic evidence of the bridge from 1773 and it was rebuilt by trustees of Cricklade Waylands Estates in 1854. The Waylands Estates, a charity begun in Elizabethan times, raised funds that helped maintain roads and river crossings. In 1692 a former structure had a twenty-two-arch causeway of timber and stone. The river here was an important highway from London during the Roman era. Historians believe that Ermine Street crossed the Thames nearby and that the Romans altered the flow of the river to prevent it flooding a causeway. The route is likely to have origins with Neolithic flint-traders from the Wiltshire Downs. This suggests that people crossed the Thames here before 2,000 BC.

This is one of many places along the Upper Thames that seems suspended in the Middle Ages. The historian Samuel Rudder said the Thames was navigable to Cricklade in 1775. Near Town Bridge was Knoll Wharf (until 1854) and moorings for barges of up to six tons. There was also a weir (possibly at SU102940) near Rose Cottage. Knoll Cottage stands as a reminder of those days before the wharf was dismantled because the T&S Canal had become a more popular route. Barge journeys downriver from Cricklade with cargoes of such commodities as bacon, barley and malt began at Horsepool, near Cricklade's old North Gate and north walls. Local legend says that wounded crusaders were brought by barge to the St John the Baptist hospice, founded nearby in the mid-thirteenth century and used by travellers for 300 years.

The Thames was a notional frontier between Saxon Wessex and Mercia, but the river had doubtless been a territorial boundary since prehistoric times. In 628

Town Bridge, Cricklade (SU102940). Cartographic evidence shows a bridge from 1773. A plaque on the bridge states 'Rebuilt by the Ffeofees (trustees) of the Cricklade Waylands Estates, AD 1854.' Knoll Wharf for barges was adjacent to Town Bridge, which carried the turnpike road between Cirencester and Swindon until a by-pass was opened.

Penda, king of Mercia, went into battle at Cirencester against the Gewissae king, Cynegils and his son Cwichelm. In that campaign the Gewissae men pushed north and crossed the Upper Thames, in all probability at Cricklade, and into the territory of the Hwicce – subjects of the Mercian king. The Hwicce were Anglians who arrived in the East Midlands in the sixth century, and later made their episcopal southern boundary between Cirencester and Kempsford. Their diocese was centred on Worcester from 680; Gloucestershire was not formed until 1007 by Edward the Confessor.

In 878 the Danes moved their 'Great Army' under Godrum to Gloucester and then re-crossed the Thames to the royal settlement at Chippenham where they caught King Alfred unawares. They set up winter camp there and like most forts it relied on proximity to navigable watercourses for transport of reinforcements. Alfred fought back the next year and drove Godrum, now a Christian with the name Athelstan, into the area around Cirencester.

The *Anglo-Saxon Chronicle* relates that in 903 Aethelwald landed in Essex, crossed the kingdom of Mercia and, having arrived at Cricklade, crossed the Thames into Braydon Forest. Another invader came in 1015; this was the Dane, Knut (perhaps better known as Canute), who landed on the south coast. He conquered Wessex, and the next year crossed the Thames at Cricklade into Mercia, which he also overpowered. He then reigned peacefully until his death in 1035.

North of Town Bridge at Cricklade is the twin-arch High Bridge (SU102941) over a less significant stream of the Thames. The 1828 OS 1in map shows it as New Bridge, which perhaps improved the former causeway crossing. Town Bridge and High Bridge carried the turnpike road between Cirencester and Swindon, up until a by-pass was opened in the autumn of 1975. Today, the old turnpike is quiet, but I sensed the spirits of Roman and Stone Age men; a crossing this ancient must surely have some tales to tell.

> *Cricklade's name origin is Old British (Celtic) and derives from* crecca-gelade *meaning 'stream-crossing by the hill'; doubtless referring to Hailstone Hill. Excavations in the parish have produced Iron Age and Roman coins and there are Roman villa remains near the old north gate. There are more Roman remains to the south-east and across fields to the north-east aerial photographs suggest the junction of five probable tracks, perhaps originally prehistoric routes.*
>
> *By the second half of the seventh century, the town was a large Saxon rectangular settlement enclosed by a defensive bank and ditch. Cricklade became a Saxon* burh *of some importance; part of a widespread system of fortified towns developed by King Alfred based largely on old Roman ones. Many were in the Thames valley 20–25 miles apart (a day's march) and intended to 'control' river movements by the Vikings.*
>
> *Cricklade church is dedicated to St Sampson, a sixth-century Celtic saint who was a missionary in Britain and a bishop in Brittany. There is a sixteenth-century four-pinnacled tower at the transepts crossing, though it has Saxon origins with Norman additions.*

> The tower vaulting has heraldic decoration, creating a strangely foreign atmosphere. The churchyard's north wall is of vertical stone slabs, as used elsewhere around the Upper Thames. Near the old north gate is St Mary's church of around 1150 with a fourteenth-century lantern-head cross in the graveyard.
>
> During the eleventh century Cricklade and Malmesbury had mints and some Saxon silver pence may be seen in Cricklade museum. King Athelstan, a great unifying power in his day, had earlier called for unification of the coinage throughout the realm with at least one mint per borough. In 1155 Henry II confirmed the town charter and Cricklade soon had regular busy markets – in fact, cattle sales only ceased during the Second World War. The town's rural past links it with cloth-making and a glove-making industry is linked with a tannery near Town Bridge; a small factory continues the craft. The turnpike roads through the town brought trade, from which the numerous inns and blacksmiths benefited.

In the east of the town Thames Lane leads to a ford (SU104938) noted in 1839 tithe records. Nearby stood Hatchetts Bridge; a rustic trestle bridge with masonry abutments, and probably named after Hatchetts Farm run by Joseph Cuss around that time. The bridge survived well into the twentieth century.

When mill leats were contructed, bridges often became necessary for continued access to farmer's fields.

Nowadays a bungalow named Hatchetts is an historic reminder. 'Hatchetts' could have derived from hatches or floodgates that were built to sustain deeper water in the river. A modern concrete bridge (SU103939) takes a track to the local cricket-ground in Hatchetts Mead, beyond which is a rivulet, Hatchetts Pyll (or creek), used to mark a boundary in the past. Both are mentioned in the fifteenth century. The ford was the venue for river baptisms, the last one being performed during the 1890s, and a contemporary photograph shows a large number of witnesses.

section two
Cricklade to Lechlade

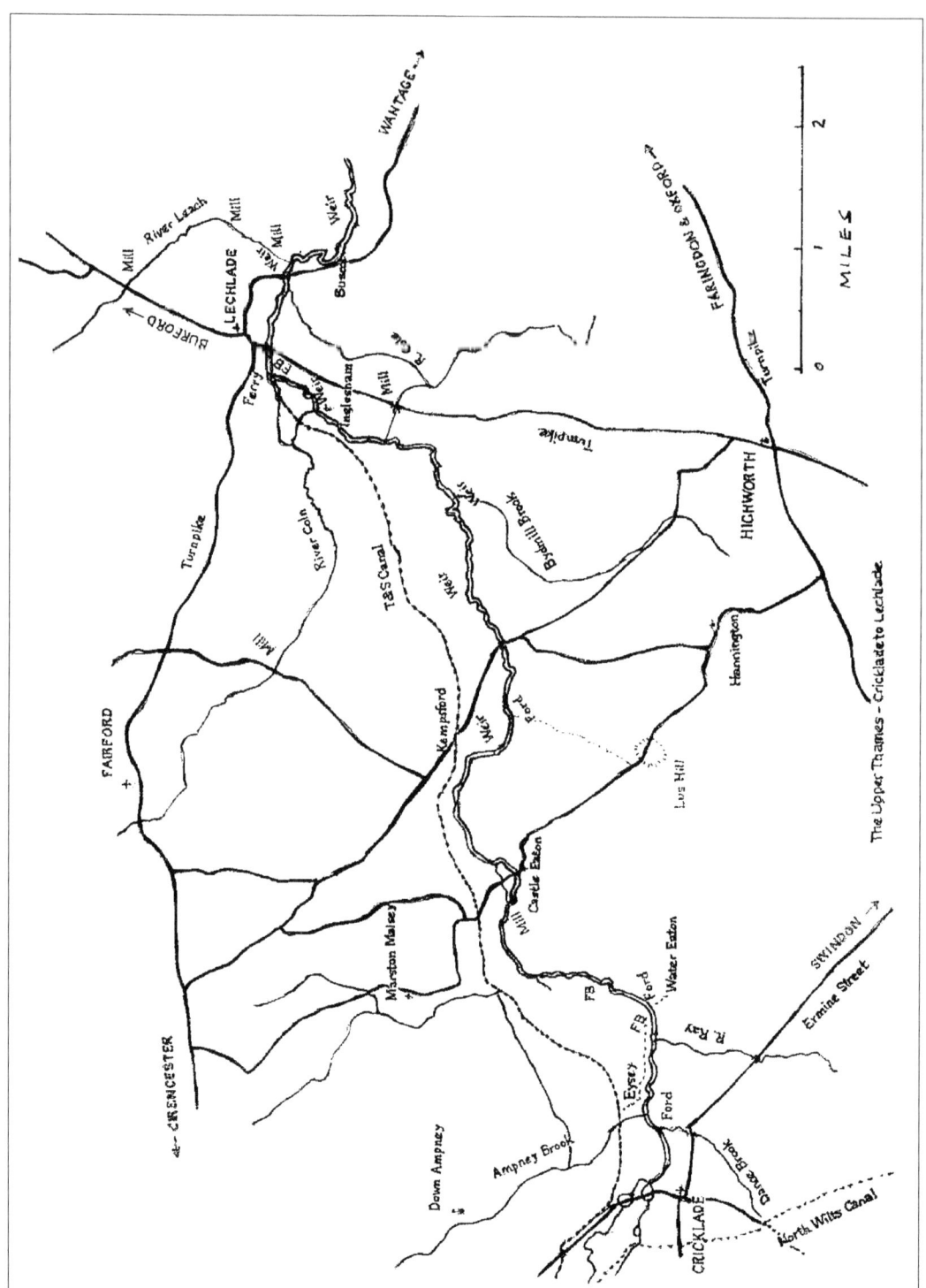

The Upper Thames – Cricklade to Lechlade (10 miles).

Cricklade to Lechlade

Since 1866 the Thames Conservators' authority and the right of navigation began at this 'official upper limit', which only the smallest of craft can now reach as the river is seldom more than a few inches deep and choked with vegetation; in fact, in 1928 the Thames dried up completely at Town Bridge. A riverside path leaves Hatchetts ford and crosses meadows to Calcutt where the meagre flow of the River Key joins the Thames from the south. In the mid-sixteenth century this was Bradenwater, or Stockum named from the tree-stumps lining its banks (perhaps pollarded willows or remnants of Braydon Forest).

> *More recently the River Key was called Dance Brook, particularly where the road from Wootton Bassett crosses Dance Common south of Cricklade. An ancient highway from the Saxon burhs of Malmesbury and Bath came into Cricklade, and by 1775 it was a turnpike road for coaches to Oxford along a route south of the Thames.*

The Thames now flows with greater vigour and passes under the ugly but functional concrete span (SU108938) carrying noisy traffic on the A419 dual carriageway, beside which some Roman remains have been found in the more rural landscape around the Thames.

East of Cricklade the Thames becomes broad between water meadows, and is often willow-lined. Most riverside willows were once pollarded producing a valuable crop of osier shoots (*rods* in the medieval era), which were deftly used by millers, lock-keepers and their families to weave such items as baskets, and traps for eels and fish like salmon. There has been a revival of osier-craft recently with the making of decorative garden features.

> *Herons find this area a productive hunting ground among the many brooklets that feed into the Thames. Young coot may sometimes be seen bobbing on the sparkling sun-lit water that can be three-quarters of a metre deep. Swans nest here in springtime and their cygnets take to the river from May onwards and learn to grub among the vegetation. There is a penalty for killing the swans; however, many have died with lead poisoning from fishing-weights. This brought a population decline, now halted following a ban on the use of lead-weights.*

Eysey Footbridge (SU112941) formerly the site of a ford shown on old maps. River navigation improvements meant many fords were replaced with footbridges.

The riverbank path crosses a footbridge over a side stream, from where a rainbow-arch concrete bridge (SU110938) over the Thames can be seen. This structure conceals a pipeline over the river's murky waters. Many years ago, according to old maps, there was a ford near this spot; today there is a concrete footbridge (SU112941) with timber handrails. River navigation improvements meant many fords were replaced with footbridges.

The Domesday entry for Eysey (*Aisi*) recorded two mills and it is possible that the mill weirs created shallows for a ford. Archaeologists say that Bronze Age and Iron Age people inhabited this area and they surely crossed the river in which they fished. Even today there are many potential crossings at gravelly places in the riverbed; yet other parts are almost choked with reeds. In bygone days the reed-cutter not only had to please millers with a flowing river but fisherman too, who wanted a habitat for fish. Most rivers and their tributaries were well managed for access to scattered settlements, even before the Roman invasion.

> *Eysey once saw regular traffic on the T&S Canal. From 1831 a lock-keeper's cottage provided lengthman's accommodation, a hay store and stabling for horses, but over the years the isolated cottage succumbed to ivy. Many canal-side buildings have surrendered to nature since the day that smoke last rose from a once-welcoming hearth. A more permanent canal relic is at Eysey Manor Farm where two canal mileposts serve as gateway markers and indicate mileages to Walbridge (west of Sapperton Tunnel) and Inglesham (near Lechlade). Walkers can explore the old canal where the former towpath is a right-of-way. From Eysey a return to Cricklade follows the path alongside the canal bed, now full of wildflower vegetation, where bird-song fills the air. Eventually the path crosses a bridge over the A419 to Wharf Farm, mentioned earlier at North Meadow.*
>
> *The T&S Canal passing through this area suffered leaks into the gravel through which it was cut. A temporary solution in 1790 was to duct water from the nearby Ampney Brook. Unfortunately this was begun without the permission of the landowner, Lord Eliot, who had the work stopped. A compromise was eventually reached; the canal would take water only between the summer and the start of autumn rains. However, the ultimate solution took place at Thames Head where a borehole was sunk and a six-sail wind-engine lifted water at several tons/minute into the canal. The wind-engine was superseded by a steam-pump, which began working in 1792. The 'borrowed' water was returned to the Thames at Inglesham after supplying the locks.*
>
> *Canal leaks were recurring troubles for the Company and problems for users of the canal and river as a highway. Early in the canal's life the engineer Robert Whitworth had its depth and width increased to ensure sufficient water for operating the locks. But this logic was flawed as it increased the vulnerability to leaks, so in the 1830s the depth of water was reduced. Then in 1879 it was reduced again but this unfortunately led to the banks drying out.*
>
> *North of Eysey lies Down Ampney whose name was recorded as* Omenei in Domesday Book, *which referred to Amma's island. This is the most southerly of the villages grouped around the Ampney Brook and some historians support the local belief it was here, on the boundary between Hwicce and West Saxon territories, that St Augustine called a meeting in 603 to urge harmony between the bishops and preachers. Tradition says they met at Augustine's Oak, mentioned by the seventh-century historian, Bede. The oak, marked on the 1828 OS 1in map, was felled in 1825 and Oak Road is now the only reminder.*
>
> *The Knights Templar built All Saints church, which dates from 1265. The crusader Nicholas de Valers, whose monument is in the Lady Chapel, was a benefactor. The church, restored*

> in 1897, has an interior dating from the thirteenth century and red flowers painted on the nave arcades. There are kneeling effigies of two Hungerford knights who died in 1634 and 1637. A memorial plaque tells us that RAF planes with gliders took off on D-Day from a nearby airfield, and another associated one at Blakehill Farm west of Cricklade, bound for Arnhem and the Rhine. There are stained-glass window memorials to the pilots and to Flight-Lieutenant David Lord, who won a posthumous Victoria Cross for his gallantry over Arnhem and who lies buried here. Blakehill airfield has now reverted to a hay meadow with a biologically diverse variety of grassland plants.
>
> Beside the church, behind an evergreen hedge, stands Down Ampney House, once the home of the powerful Hungerford family. The stone house of Tudor design built for Sir Anthony was renovated in 1799; partly to plans by Sir John Soane, noted architect of the now much-altered Bank of England. The Hungerfords were very active in this neighbourhood, and some were buried at St Mary's church, Cricklade. The Old Vicarage at Down Ampney was the birthplace in 1872 of the composer Ralph Vaughan-Williams when his father was vicar here. Vaughan-Williams composed the tune 'Down Ampney' now used for the hymn 'Come Down, o Love Divine'.

The Thames is strengthened by the confluence with the River Ray, known as the Wurf from the ninth century until sometime in the 1600s, the name deriving from its twisting, turning nature. Near Water House there is a plank footbridge (SU124939) with stone abutments, once known as Horse Bridge and perhaps a former crossing place for farmer's horses. In 1307 there was a ford and crossing the bed of firm gravel would have been easy going but, judging by the debris lodged about 4ft (1.5m) up the bridge supports, crossing would sometimes have been hazardous. Pausing on the bridge and gazing into the clear water, minnows can be seen weaving in and out of the shadows. County maps by Christopher Saxton (1577) and John Speed (1610) show the crossing linked the hamlets of Isey and Nunyeton, both places of pre-Conquest origin with two mills recorded in the Domesday Survey. The latter village belonged to Godstow Priory, near Oxford, in around 1281 and had a mill-house near Water Eaton Weir in 1535. The weir survived until 1886.

The river begins a rather isolated and inaccessible stretch flowing northwards soon coming to the site of Cow Neck. This was once a large side stream on the western bank, of which little remains except some riverside pools. The riverbanks are thick with yellowing reeds and tall sedges, still bearing the tattered remains of last summer's blooms. Reeds cultivated from early medieval times were used for such rural crafts as weaving all kinds of lightweight containers, but also used for roofing and as fuel.

A rivulet from Marston Meysey joins the Thames. In the early thirteenth century Robert and Roger de Meisi held land here. West of the river crop-marks have been seen near a track once known as Ridgeway Lane. Aerial photographs show crop-marks of probable tracks heading for the river, suggesting an ancient ford existed in this vicinity (perhaps at SU135960). The ford could have served a drove-road (or perhaps portway), passing old Port Farm and going south towards Swindon.

The T&S Canal ran very close to the river here: a humpback bridge and a watchman's roundhouse (both private property) remain beside a field path to Marston Meysey. At the height of canal activity there was a place for barges to lie clear of the channel, and a wharf near the roundhouse where goods would have been discharged or loaded.

The next stretch of river forms the Wiltshire-Gloucestershire border established in late Saxon times. The river divides (SU140960) but the boundary follows the

A girder-&-plank footbridge (SU124939) with stone abutments was once known as Horse Bridge. The crossing linked the medieval hamlets of Isey and Nunyeton.

weaker northern course flowing through marshy land and crossed by a small iron handrail bridge (SU143967). The southern loop fed a leat for a busy mill, which existed from at least 1790 until the late nineteenth century and is on the 1828 OS 1in map. Following along the old leat, and disturbing a heron, brings us to the foot of Mill Lane, Castle Eaton, where a footbridge (SU144958) for the Thames Path crosses above an old channel shaped in stone.

In this area are fields named Black Gore in 1763 and Borstead in the mid-1500s. Archaeologists say that place names that include 'black' can indicate habitation sites. They may have been part of a recorded *burh* of unknown location. Could Blackburr Farm (SU142964) be the *burh* site north of the iron-rail bridge (SU143967) over the former northern stream?

The stronger stream flows close to Castle Eaton and is crossed by a rather ugly cast-iron bridge (SU144958), reminiscent of those used for railways. It was made by iron-founders Finch & Co. of Chepstow and installed by the Thames Conservators, but the structure replaced an older timber bridge with handrails carried on out-riggers and supported on five stone piers. Early twentieth-century photographs of the old bridge are in the adjacent Red Lion Inn and show the road on a raised section of some 40m additional to the bridge. In the 1690s it is said there was a long causeway of about thirty timber arches and a twin-arch timber bridge on stone piers. County maps by both Speed (1610) and Saxton (1577) show the bridge carried the road from Hannington to Fairford, which both began as Saxon settlements.

Domesday Book describes Ettone *as 'a river-side farmstead' held by Turold, a tenant of Earl Roger of Montgomery. Not until the fifteenth century did the prefix 'castle' come into use. It is a village of Saxon origin with some stone-built cottages. St Mary's church is largely the result of work by W. Butterfield in 1863. The ancient tower and an open bell turret overlook the landscape from a grassy bank by the river, which flows quietly through the meadows.*

A remnant of a drove-way runs south near Lus Hill where Henry atte Drove resided in 1332, maybe at Droveway Farm. In Domesday Book Lus Hill was Rustesselle *with two mills in the parish. Lus Hill's name is thought to derive from Old English for spindle, a common tree in these parts and the source of spindles used when making woollen yarn. A track, known since the 1420s as Stapler's Lane (doubtless used by the English wool staplers), continues southwards beyond Hannington. Presumably this was a regular route of drovers and wool-merchants. An old rural song says 'the drover's life has pleasures the townsman never*

The 'old' Castle Eaton Bridge (SU144958) survived into the early twentieth century; the crossing was on a fourteenth-century drove-way route. (By kind permission of Mr & Mrs Lyall, the Red Lion Inn)

> knows', which may ring true on idyllic summer days though less so in winter. The Cotswolds were grazed by vast flocks of sheep largely belonging to religious communities. The high-quality wool was an important export via Southampton, mostly to Italy. But the trade was not all one way; canvas was imported to places such as Chipping Campden for making woolsacks. Until the nineteenth century, the majority of the inhabitants of Hannington were still employed in agriculture.

The next town along the river is Kempsford, whose Old English name was *Cynemaere's ford* – a crossing (perhaps at SU161963) – and most probably going into Blackford Lane. In the early fourteenth century John de Blackford lived at Blackford Farm (just south of the river), and aerial photographs show cropmarks east of the farm, which include possible trackways on a north–south axis seeming to pre-date the medieval ploughing ridge-&-furrow pattern. The ford is allegedly haunted by the ghost of the drowned son of Henry, earl of Lancaster, who lived at a nearby castle in the fourteenth century. The Conservancy closed the ford in 1869, but the stone bed was still there in the early 1900s when an almost dry-shod crossing could be made, most probably because the river was about 20ft (7m) wide.

A Saxon ring-brooch of the fourth/fifth century was found near the ford and further along the river a Bronze Age spearhead was discovered, thought to have been committed to the river in a peace ritual, perhaps after a boundary dispute. The Thames has one of the highest densities of prehistoric weapon deposits and it is thought that some stretches were specially revered. Although many weapons are found near old crossings, and possibly deposited to 'protect' the crossings, shallow water is the most likely place for finds. Other items may have been committed to deeper water and have not yet been discovered.

Most rivers once teemed with fish and the Thames was no exception; a field named Fishery Eyot (island) (SU155966) on land owned by Lord Coleraine appears on the 1802 Enclosure map. That map also shows a weir (SU163963), doubtless once a fish-weir that was removed by the Conservancy in 1869. An adjacent field was Gullet Meadow – a 'gull' being shallow swift water below a weir. The county boundary follows a lesser stream that was the main one in 1802 until the early 1900s. The *Anglo-Saxon Chronicle* relates that Aethelmund, a Hwicce leader, crossed the much-contested Wessex-Mercia boundary (the river) in 802 to do battle with Weohstan, a West Saxon prince. Both were slain but Wessex took the victory. In 825 another battle due south at Wroughton led to a victory for Ecgberht, king of the West Saxons, thus ending Mercian dominance in the region.

The Domesday entry for Kempsford mentions four mills (some doubtless on the River Coln), meadows for hay and sheep-grazing pastures (the same applies today) and cheese production (most likely from ewe's milk). Near the river are numerous drainage channels possibly once used for 'drowning' hay meadows to protect them from frost and to promote early summer grass. Hay for the once vast population of horses in the countryside and the cities was harvested from water meadows and loaded onto Thames barges bound for Oxford, London and elsewhere.

Kempsford's church of St Mary the Virgin, begun in 1090, has a central tower built later by John of Gaunt and with weathervanes on all four pinnacles. The interior of the tower is vaulted and painted with shields-of-arms and red roses of the Earls of Lancaster who held the manor from about 1300. There is a memorial brass, with a merchant's mark, to Walter Hichman who died in 1521. The brass shows him in civilian dress with his wife Cristyan and their four sons. The chancel was enlarged in 1855 when G.E. Street's restoration was carried out. He also designed the choir stalls with their leaf-fan finials. Most of the stained glass is also Victorian.

A castle once stood on the riverbank near the church, and at one time it belonged to Blanche Plantagenet (patroness of Geoffrey Chaucer). Blanche married John of Gaunt (from Ghent) in 1359. Manor Farm now stands on the same spot; it dates from the sixteenth century and was rebuilt in 1846. After Henry VIII disbanded the monasteries, Kempsford manor lands were granted in 1549 to Sir John Thynne. In the seventeenth century Thomas Thynne built a mansion at Kempsford, but the Coleraine family had it demolished in 1784, and much of the fabric was purchased by Mr Lovenden to build Buscot House.

At Kempsford, the T&S Canal Company owned a wharf of which traces are mostly lost; however, a canal agent's house dating from the late 1780s is in a good state of preservation. There was a swing-bridge over the canal for access to Thames Meadow and to the river only 50 yards away. The north of Kempsford parish is dominated by the massive Fairford airfield begun in the Second World War and used as a United States Air Force bomber base for ten years from 1953, and later for Concorde pilot training.

Kempsford is linked to Hannington and Highworth across Hannington Bridge (SU175961), which marks the modern limit for river-craft with a draught of no more than three quarters of a metre. Nevertheless, my attempt by dinghy from Lechlade proved quite a struggle with only two oars-power negotiating the shallows and the weeds swinging in the current. This may compare with the state of the river at the end of the nineteenth century when it was said in an assessment of the decline of navigation to be 'heavily overgrown'.

This is a lonely stretch with only an occasional meadow 'beach' where cattle come to drink and where the otherwise sheltered water becomes open to any movement of air. The river snakes so tightly it seems to close in as each new scene opens ahead. The riverbank vegetation casts deep shadows and very little can be seen from a small boat, and in some parts the river is invaded by young willow saplings. On this particular spring day snow flurries and an icy wind whipped up small whitecaps to mark my return past Inglesham. A sense of 'exploration' crept into my being. Perhaps the very first immigrant explorers venturing up the river experienced similar feelings.

The present-day three-arch stone-built Hannington Bridge bears the date 1841 and it replaced a predecessor of timber on stone piers begun in 1647 – 'because the thirteenth-century bridge was damaged in the Civil War'. There was once a causeway of sixteen arches for pedestrians but wheeled traffic had to go through

Hannington Bridge (SU175961), 1841, the current head of navigation for craft with a draught of no more than three quarters of a metre. Kempsford and Hannington are linked by Hannington Bridge.

the river – suggesting a bridge and a ford co-existed for many years. Some historians claim there was a Roman bridge here; archaeologists have found a scatter of Romano-British debris in a meadow just to the north and the remains of a Roman villa to the south-east. In 1281 Robert de la Brigge had Bridge Farm, and in 1333 Alan atte Brugge lived close to the river. Saxton's 1577 county map shows the crossing, which only twenty years later was recorded as Thames Bridge. Later still, there was an alehouse on the southern bank.

> Beyond Hannington lies the quiet but once important market town of Highworth. Its origin dates from pre-Conquest times when it was called simply **Worth**, *meaning 'a small settlement'*.

> *The prefix 'High' was added during the thirteenth century when a market was established. It is well named though, with distant views all around – especially from the church tower, which like many others became a lookout post during the Civil War when the town was loyal to King Charles. In the Second World War Highworth became a military control centre for a line of defence against enemy parachute landings, which included many pillbox gun-positions along the Thames. Plans were made to flood meadows in the vicinity of Castle Eaton, with the intention of protecting from parachute attack the RAF stations scattered through the Upper Thames area. The river became a thin 'blue' line.*

Half a mile downstream from Hannington Bridge was Ham Weir (SU182965) on the 1802 Enclosure map near Lower Ham Field, and noted again in 1865 and 1871. It was dismantled in about 1895 and all traces were gone by 1910. In the meadows north of the river near Manor Ham Barn a significant quantity of crop-marks form long linear features that may relate to ancient trackways suggesting a settlement of uncertain date. Nevertheless, one might speculate that they were Saxon, as it is known that the Saxons frequently cultivated land on the gravel river terraces, which occur in this area. However, drainage channels are another possible cause of the crop-marks. North of the river the land was once farmed by the medieval Godstow Nunnery near Oxford.

Within about a mile another weir (SU194967) (1802 Enclosure map again) was beside Kempsford Mead. Perhaps the weir was for fish (*a fiswere* was noted in the fourteenth century) possibly connected with Sterts Farm, which was part of Longleat Priory estate and leased from Lacock Abbey by William atte Sterte in 1327. Near here Bydemill Brook (driving a mill in 1317) joins the Thames. South of the river is a Roman villa site – confirmation of enduring habitation close to the river highway.

The Thames now turns sharply north-east towards Inglesham through what some might term dull country; its sinuous course is a highway hidden by overhanging old trees where trout favour the stretches of shaded water. An old course of the river Cole, once called the *Lente* or *Lynte*, possibly from Celtic *lliant* meaning 'flood-water', is a right bank tributary. The name survives at Lynt Bridge where Lambourne Lane (bridleway) leads to St John's Bridge at Lechlade.

> Until 1854 the Cole defined Wiltshire's boundary with Berkshire; the same as in 854 when the Cole was the Smite – thought to mean 'gliding'. The river's headwaters were utilised as a summit supply for the Wilts & Berks Canal. At the end of the twentieth century, an experimental stretch of the Cole was restored to a wild state with curves and gravel beaches re-created to slow the current, which may alter the impact of floodwaters, and attract wildlife.
>
> Less than half a mile up the Cole was a mill (existing in 1761) near a small bridge. A 1773 map of Wiltshire by John Andrews and Andrew Drury shows the mill and a weir, noted again in the 1790s. The bridge was Inglesham Bridge referred to in a 1619 document, and described as stone-built in 1675 by John Ogilby, cosmographer to Charles II. It carried a Salt Way, later a turnpike route (now the A 361) from Burford through Lechlade to Highworth, and beyond to Salisbury and the south coast. Salt ways began in Saxon times, but were much used by general traffic. The salt was a very important commodity for preserving fish and meat, but also used in personal hygiene.

Almost opposite Inglesham church there used to be Redpool (SU204983), maybe referring to a 'reed pool'; many reed-beds existed locally. A little further on there was Inglesham Weir (SU205985) – shown on the 1839 Enclosure map – where Elizabeth Golding owned a field named The Wire Ground. The weir was removed in 1868 after complaints about it obstructing navigation. There is a small, almost dry stream on the right of the Thames below the site of the weir. This was Murdock Ditch and used to define the county boundary, though why an apparently insignificant channel was used is a mystery.

> In 950, Inglesham was Ingenesham, meaning 'Ingin's water-meadow'. In fields between the Thames and a branch of the River Cole are the hummocky remains of a medieval village. Rural settlements experienced many changes in the 1870s when agriculture suffered a depression and thousands of workers left the countryside in favour of work in towns. The fields would once have been worked intensively to provide cereals and root crops. A meagre diet of bread and potatoes for many nineteenth-century farm-workers' families would have been readily supplemented from the countryside with mushrooms, watercress, nuts and berries; not to mention the occasional rabbit or even a bird's egg. Today Inglesham is little more than a few houses, and a farm, which was once a priory.
>
> The tiny church of St John the Baptist originated about 950, it stands close to the river. At one end is a small cote with two bells. This ancient building mostly dating from the thirteenth century has a melancholic air. There are old wrought-iron hinges on its door, uneven floors

A rustic footbridge near Inglesham.

> inside and wall paintings; a seventeenth-century pulpit and box-pews; but the Saxon carving of the Virgin and Child is a gem. The church was repaired in 1889-90, but saved from over restoration or falsification by the pre-Raphaelite poet and artist, William Morris.

North of Inglesham the River Coln, the *Cunuglae* in Saxon times, enters the Thames from the north-west and all at once the Thames is deeper and wider. It has fallen nearly 150ft (50m) in the 20-odd miles from the source. This confluence (SU205987) was the much-debated spot chosen for the T&S Canal junction with the river, the location only being confirmed just before being cut in May 1789. Inglesham Lock controlled a fall of 6ft (2m) into the Thames, and the canal was opened to traffic later the same year with celebratory cannons being fired from nearby Buscot Park, owned by Edward Lovenden.

The bridge over the lock bears a date confirming the canal's completion in November 1789. However, by 1813, due to Inglesham's relatively isolated location, the canal-side wharf here was closed and the Canal Company purchased Parkend Wharf at Lechlade. Near the former canal junction and the watchman's roundhouse, a footbridge (SU206989) over the Thames replaces the earlier timber bridge (SU205987), known as Donkey Bridge of which part of the abutments remain. It was gated at one end to prevent livestock entering, and it took the towpath south of the river as far as St John's Lock. In 1839 withy beds here belonged to Benjamin Hodges, who lived in St John's Lock cottage. Since the sixteenth century reeds and rushes had been harvested and stored at Lechlade Wharf before being shipped to local chair-making establishments and various other industries.

Lechlade has many Georgian houses, a few old coaching-inns and some eighteenth-century 'listed' buildings, but archaeologists have found traces of a late Neolithic farming settlement to the north-west. The finding of a Stone Age hand-axe indicates the earliest human activity. There was also an Iron Age village nearby, which was taken over by the Romans, and then Saxons established a settlement in the late fifth century. The Welsh Way, a drovers' road, passes through Lechlade en route to Wantage and London, much used by dusty traders with livestock.

The town grew in status during the Middle Ages; it had an annual fair on St Laurence's Day (10 August) from 1210 onwards. In the 1700s, the fair was held near the Thames, but with floods in 1774 it was held ever after within the town. The Tuesday market sold quantities of cheeses; in more recent times the town bustled with regular livestock markets and horse-fairs where animals, their harnesses and trappings could be bought. With the arrival of the railway, the market was held closer to the station. By way of recreation, a late summer river carnival attracted crowds of spectators during the early twentieth century.

Daniel Defoe visited Lechlade in the early 1720s (before the T&S Canal was built) and described the town as 'at the Head of Navigation with large barges at the quay'. There was a well-established 'free' wharf owned by the town's officers and run by Richard Gearing, a coal and corn merchant, until William Hill, a Cirencester merchant acquired it. In those days river trade was at its height; wharves were busy with waggon traffic to and from local market towns, merchants with agricultural produce and Cotswold clothiers sending goods on barges to London and many places beyond.

Parkend Wharf and one east of the town with a total length of almost 180ft (60m) permitted loading for barges plying the Thames to the capital city. Parkend Wharf handled timber, hay, coal and grain. A cheese warehouse and salt store stood close by; flour and butter were regularly shipped, as well as Cotswold stone. From the early 1800s, the wharf-side premises of Matthew Hicks handled corn, salt and hay, and his son continued trading as a maltster, but from about

A Thames & Severn Canal roundhouse at Inglesham (SU205987), where the canal joins the Thames. When opened, the canal brought increased river traffic affecting some Upper Thames crossings. Many fords were grubbed out and numerous bridges had to be altered. (Courtesy of www.old-england.com)

> *1873 coal was sold at the railway station. The railways almost killed river traffic reducing it mainly to bulky goods like timber and stone. Near the church was a malthouse owned, at one time, by Elizabeth Golding, who had also owned the older Wharf Ground. Her family had long-established ties with the river.*

Adjacent to the former Parkend Wharf is Ha'penny (originally Town) Bridge (SU213993), so named from the original toll for pedestrians, levied until 1839, though it was raised from ½d to 2d. In 1792 the Burford to Highworth road was upgraded to a turnpike route and Town Bridge (probably designed by Daniel Harris of Oxford) replaced a ford and ferry. The bridge builders used bundles of faggots in the bridge foundations, which were exposed during repairs in the mid-twentieth century and found to be still in good condition. The bridge tollhouse remains intact, having been inhabited until recent times.

The ford (SU212993) was approached via Tidford Lane (now Bell Lane, named after the Bell Inn) near the free wharf. The ferry crossing – the highest one on the river – was a short distance up river from the later bridge. The ford was dredged away in the 1960s, though the entrance to the ford remains. Maybe the river was shallower then too, which could explain why it froze in 1963 to the delight of many ice-skaters. Even in the sixteenth century the Thames froze and the first 'frost fair' was held.

South of Ha'penny Bridge the former towpath leads to St John's Lock (SU222990), built originally of timber, which included a wharf in 1784. This was the first lock after the T&S Canal. In 1793 the lock toll was 2½d (1.25p) per ton, and in 1830 a two-storey lock-keeper's cottage was built on an island (owned by Benjamin Hodges) created by the lock channel. The weir was repaired in 1867 and the lock rebuilt in 1905. A new single-storey keeper's cottage was built south of the lock in the late nineteenth century. Beside the lock is Raffaelle Monti's reclining figure of 'Father' Thames who casts his paternal eye over river traffic. Here, too, is the meeting of Gloucestershire with Oxfordshire.

> *The River Leach flows into the Thames near St John's Bridge. Lechlade.* Lecelade *in the eighth century, is thought to mean 'crossing of a streamlet in boggy land'. The Domesday entry records 'meadows for hay; a fishery yielding eels [begun in Saxon times]; and three mills'.*

Ha'penny Bridge, Lechlade (SU213993), named from the toll for pedestrians, which was levied until 1839, though raised to 2d. The tollhouse was a residence until recent times. In 1792 this Burford to Highworth road was up-graded to a turnpike route

The entrance to Tidford, Lechlade, is at the foot of Bell Lane (SU212993). The ford was dredged away in the 1960s. The ford continued in use after Ha'penny Bridge opened

Probably the most unusual Thames bridge, albeit temporary. In the autumn of 1909, troops on manoeuvres used Polyansky floats and planks to cross the Thames. (Courtesy of Lechlade History Society, from their archive collection)

> *Priory (or Lower) Mill and Lade Mill were both on the Leach a short distance from St John's Bridge. Lade Mill worked from the thirteenth century and in 1839 it was 'Giles's Mill', run by John Giles. In the eighteenth century Lower Mill was owned by a Thames barge-master who traded in grain.*
>
> *In about 1245, Isobel de Mortimer was given Lade Mill and 'The Lade' (where there would have been a ford, across the River Leach). She founded a nunnery nearby, which Richard, earl of Cornwall (brother of Henry III), enlarged in about 1252; it later became a hospice and eventually an Augustinian Priory and a medieval travellers' rest. The Priory, dedicated to St John the Baptist, later became the Trout Inn. Before 1704 the inn was called John the Baptist's Head and later the innkeeper was Ben Hodges, who owned land at the lock.*

A little downstream from St John's Lock is St John's Bridge (SU224990). Peter FitzHerbert, husband of Isobel de Mortimer, began the original, a three-arch stone bridge with a multi-arch causeway on the approach from the south, in 1229. From about 1234 onwards a Bridge Fair took place. This event lasted some five days at the end of each August. The bridge had to be repaired in 1341 and again in 1387, and when John Leland visited in the early sixteenth century he stated that the priory had 'fallen into ruins' and that some of the 'stone recently used to repair the bridge'. The bridge served traffic between Burford and Swindon or Faringdon. John Ogilby's *Routes of a Hundred Roads* identifies a stone-built bridge in the 1670s for the longer-distance Chipping Campden to Salisbury route. Turnpike traffic used the bridge from 1727, and it was largely rebuilt in stone with a single arch in the 1830s. A bridge toll was levied on boats passing beneath; between 1751 and 1771 it was 3d (1.5p) per 5 ton for barges. Bowden's 1775 map shows an earlier flash-weir, which remained below St John's Bridge until the mid-1920s.

> *Church Path or Bridge Walk leads back to the town and to St Laurence's church whose pinnacled tower is topped with a spire and golden cockerel. The church spire is a landmark visible from the river and the clock chimes may be clearly heard. The saint was gruesomely martyred by being roasted alive in 258. The church was rebuilt in 1473, with additions in the sixteenth century. The exterior of this 'wool' church has many gargoyles and heads carved from Cotswold stone. In the chancel there are roof-bosses holding heraldic shields. A chandelier*

St John's Lock (SU222990) c.1902. (Reproduced with permission of English Heritage, NMR)

Hay harvesting at Lechlade. Hay was a valuable crop – fuel for draught horses. (Courtesy of www.old-england.com)

A view downstream from St John's Lock, Lechlade.

was donated in 1730 by Robert Ainge who was a wharf owner. In the north aisle there is a 1450 memorial brass to a wool-merchant, John Townsend and his wife. He is known to have exported wool worth over £1,000 to foreign merchants and may well have been a church benefactor. Outside the north door is an ancient penance stone and Shelley's Walk with a commemorative plaque recalling his poem, 'In Lechlade Churchyard', written in 1815.

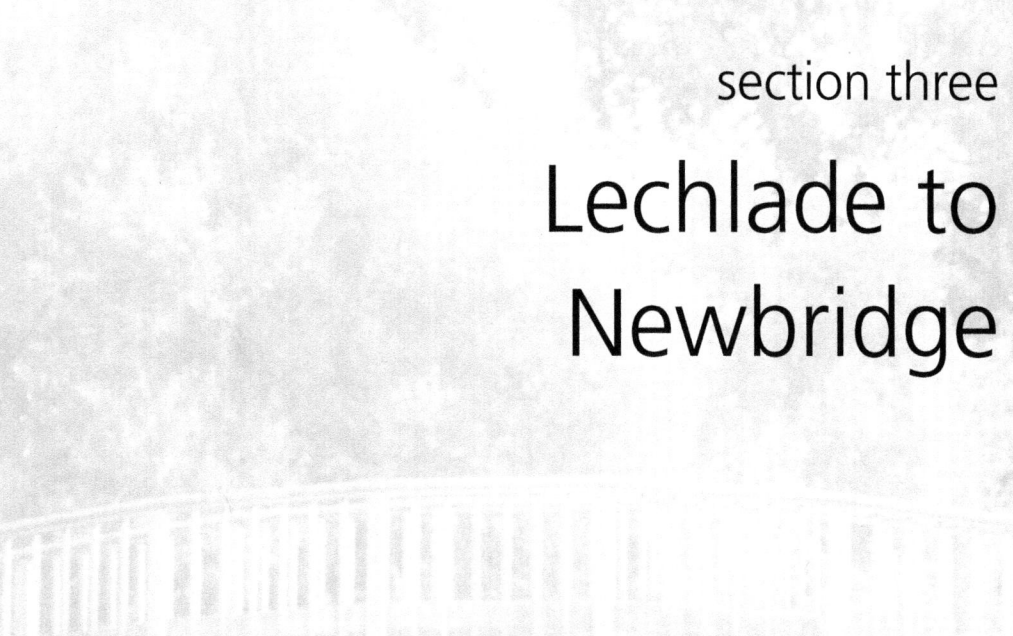

section three

Lechlade to Newbridge

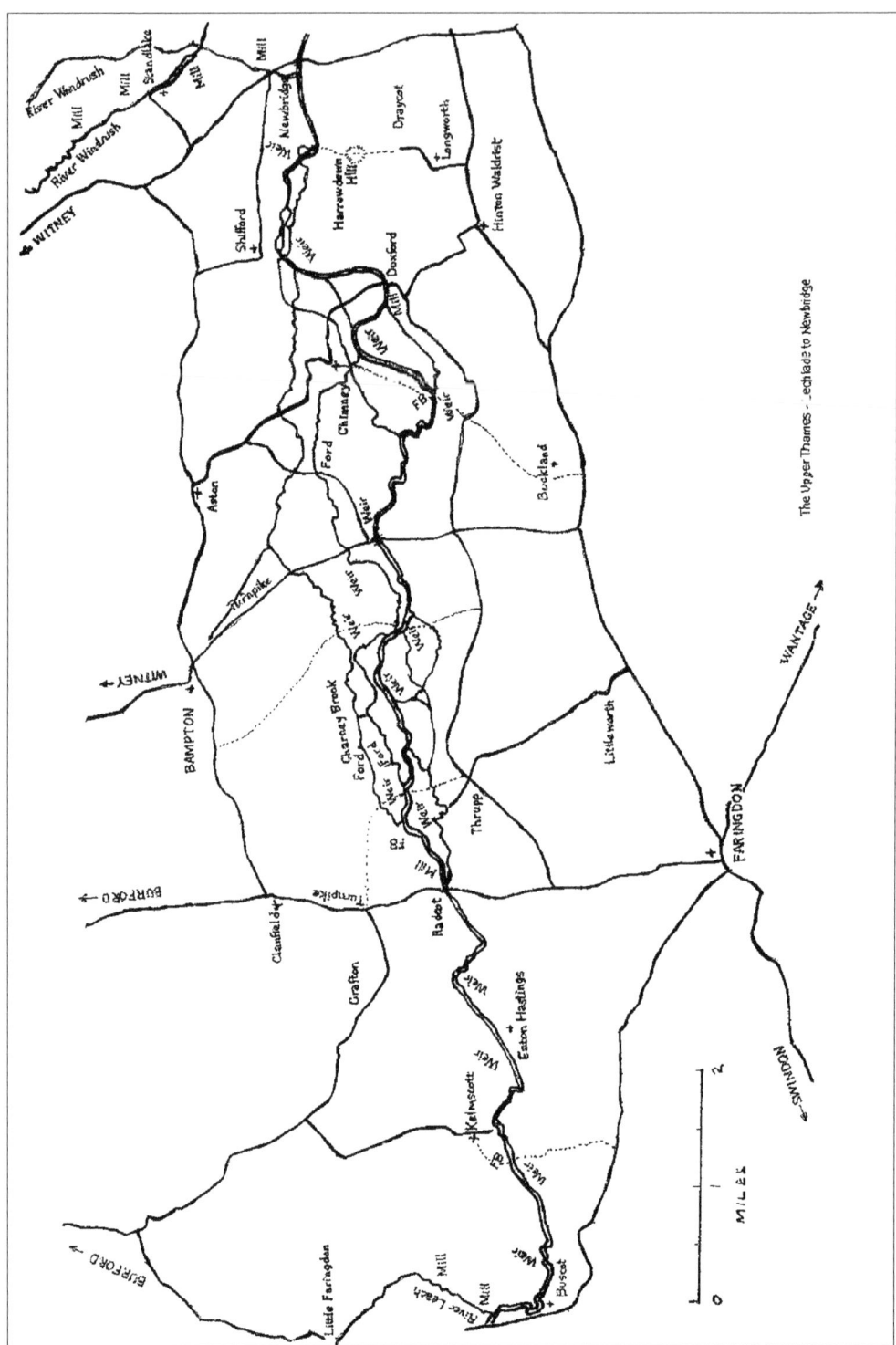

The Upper Thames – Lechlade to Newbridge (17 miles).

St John's Bridge & Weir (SU222990) *c.*1902. The first bridge was built *c.*1229 and was repaired many times; the flash-weir appears on Bowden's 1775 map. (Reproduced with permission of English Heritage, NMR)

Lechlade to Newbridge

Beyond Lechlade the Thames enters Oxfordshire, once a heavily wooded county. Wychwood Forest, probably derived from Anglo-Saxon Hwicce, extended into Gloucestershire and part of Warwickshire. The Romans cleared some forest, but later it regenerated and was enlarged by Norman kings for hunting deer and boar. Forest covered about two thirds of Oxfordshire when King John came to the throne, but hundreds of acres of woodland, much valued for foraging swine, were given up around Witney. Edward the Elder set the shire boundary in the tenth century and the road network was largely established before the Conquest. By the eighteenth century many fields produced quantities of barley, which went to brewhouses, and wheat milled in numerous riverside mills.

Some Oxfordshire rivers flow from the Cotswolds into the Upper Thames, but none of any significance from the south due to the proximity of a narrow ridge of limestone hills. The River

Windrush is believed to take its name from Old British, probably meaning 'white marsh'; later confirmed as Uuenrisc in a 779 Saxon charter. Thomas Moule's county map (1830) shows it as Wainrush where it joins the Thames at Newbridge. The River Evenlode was called the Bladene from the seventh to fifteenth centuries. These river valleys support productive farmlands and meadows where sheep grazed the limestone uplands before the Romans came. Some thirteenth-century wool-merchants became very wealthy and financed the construction of bridges on trade routes.

Most early settlements avoided heavy clay land close to the Thames and any fords would have taken advantage of alluvial gravel deposits or where the river current was reduced by meanders. West Saxon kings ruled south of the Middle Thames; according to Abingdon Abbey chronicles the Wessex-Mercia frontier was along the Thames. Until 1973 the river defined the boundary (with Berkshire) eastwards as far as Oxford, but nowadays it is further south.

The Normans began many of the major settlements; others with names of Saxon origin suggest continuous occupation from the Roman era and earlier — perhaps when prehistoric farmers settled permanently. Late in the Middle Ages agricultural activity and widespread trading became mutually supportive and there were at least twenty market towns. The change

The timber-arch footbridge near Bloomers Hole, Lechlade (SU226988), crosses the river where a ferry possibly once did.

> *from open-field farming only began in the 1750s and the final Act of Enclosure was passed in the 1880s, with benefits intended mainly for farming and crops.*

After St John's Bridge the old towpath followed the left bank as far as Radcot. The river continues its peaceful progress, making a sinuous, silvery ribbon through meadows. The course is rather capricious, flowing through a succession of lonely landscapes and passing only the occasional riverbank dwelling. It passes the curiously named Bloomers Hole and its man-made channel (1802 Enclosure map). In Bloomers Meadow (1802 map again) crop-marks have been noted, possibly tracks of uncertain date. Whether they relate to a track to an earlier ferry or weir is mere speculation, but an elegant arched footbridge (SU226988) now crosses the river, possibly where a ferry once did.

Moving on, we pass the old Cheese Wharf and a Second World War pillbox, and then soon hear water gushing over the weirs at Buscot Lock (SU229981) which was completed in 1790, though as early as 1771 a toll was taken at a flash-weir. In all probability, the flash-weir, which existed for many years, possibly started as a fish-weir. A fishery and about 300 acres of meadows were recorded in the

Opposite the old Cheese Wharf, Lechlade, is one of the many Second World War pillboxes built alongside the Upper Thames (SU225984).

Domesday Survey. In the late 1790s, a lock cottage was built and the lock-keeper was Elizabeth Gearing, doubtless a relative of Richard Gearing who ran the Free Wharf in Lechlade.

For a long time Buscot Lock was the property of the landowner at nearby Buscot Park, Edward Lovenden who was MP for Abingdon and a Thames Commissioner. He imposed a toll of 1*s* (5p) per 5-ton of river craft but also, uniquely, took a similar toll on returning craft. Lovenden was a great supporter of improvements to river navigation but was a fierce opponent of any proposed canals that might take water from the Thames and therefore jeopardise his income. In 1867 Buscot Lock was fitted with new gates but they lasted only ten years. Lord Faringdon of Buscot Park built a new lock in the early years of the twentieth century, and more work was completed in the late twentieth century. Today's lock has two weirs, the second only constructed in 1979.

> *On the adjacent Brandy Island there used to be, until the mid-1930s, a large distillery where sugar and alcohol were produced from sugar beet farmed locally and irrigated with water lifted from the Thames by a large waterwheel near the lock. This enterprise was begun around 1870 when the squire, Mr Robert Campbell, began exporting to France his contribution to their brandy industry. However, the venture had failed by 1879.*
>
> *The picturesque stone-built village of Buscot is not far from the lock; it originated in the Saxon era as a cluster of buildings named* Burg-weard's cot with a river fishery. *Prior to the arrival of William of Normandy, King Harold held the land, which William then gave to his half-sister's son, Hugh of Chester, who then let it to a tenant, Roger. In 1639 Sir Edward Yate of Buckland held the manor; his grandson married into the Throckmorton family who will be remembered for giving their name to a street in London's city acre.*
>
> *Buscot Park, created in 1780 by Edward Lovenden Townsend, has an Italianate water-garden with a 22-acre lake laid out by Harold Peto for Lord Faringdon. Buscot House is home to the Faringdon Collection of many famous works of art, in particular* The Briar Rose *painted in 1889 by the pre-Raphaelite artist Sir Edward Burne-Jones, a close friend of William Morris who lived at nearby Kelmscott Manor. St Mary the Virgin church, a short distance west of the village, has two stained-glass windows by Burne-Jones and a painted pulpit. The Cotswold stone rectory close by is in Queen Anne style.*

A rustic footbridge (SU247985) near Eaton Hastings is on the site of Hart's flash-weir (1746–1937).

Downstream where a field named Wharf Ground is shown on a 1744 map, there was a wharf (SU236979) at the head of Buscot Pill (a creek), doubtless the scene of much activity with shipment of goods. On the northern banks of the river there are archaeological traces of a settlement, possibly dating to the Roman era, and adjacent are some Second World War pillboxes – already mellowing historic structures and true examples of 1940s wartime architecture.

From Buscot Lock footpaths lead to Kelmscott and to Eaton Weir rustic footbridge (SU247985). The bridge is at the site of Hart's flash-weir (1746). Rocque's 1760s survey shows a 'wire' and Davis in 1797 shows a weir, which was reconstructed in 1791. Eaton Weir was mentioned again between 1802 and 1879 when its keeper was Mr or Mrs Hart, and it continued to exist until 1937. Although the weir had a fall of only 18in, a boat-slide was constructed in 1911 for small pleasure craft. This system helped to sustain the river level.

A twelfth-century document refers to *copman-ford* (for chapmen or merchants), possibly located below a fish-weir originating in the eleventh century when fisheries were recorded. The ford remained in use until the seventeenth century. There must also have been a ferry as a Ferry House is shown on old maps. Later maps show the Anchor inn, which burnt down in 1979 with the tragic loss of three lives. The inn had been the weir-keeper's house for many years. The innkeeper in 1883 was Mr Hart, succeeded by Mr

New in 1888. Perhaps there are ghosts of long-gone travellers lingering at this crossing.

> *Kelmscott was* Cenhelm's Cot *in 1234 and is now a straggling rural community in remote farmlands and willow-lined water meadows. Kelmscott Manor is a modest stone-built Elizabethan house of about 1570 and lies apart from the village beside a track down to riverside moorings. In 1871 the pre-Raphaelite craftsman, poet and social reformer William Morris rented the house and lived there with his family. A walk through the village passes cottages with vertical stone slab garden walls, the seventeenth-century Manor Inn (formerly the Plough), a memorial cross at the village centre and the Morris Memorial Hall.*
>
> *St George's church, dating from 1190, has fragments of medieval stained-glass and wall paintings in the north transept. When William Morris died in London in 1896 his body was brought by train to Lechlade and transferred to Kelmscott. He made his last journey, as many Cotswold farmers did, carried in the traditional manner in a farm waggon to the churchyard. His remains lie in the south-east corner beneath a lichen-encrusted stone resembling an upturned boat, which has a carved design by Philip Webb, another member of the Arts and Crafts Movement that William Morris founded.*

Between Buscot and Eaton, a little upstream of Eaton Weir, there used to be Farmer's flash-weir, built before 1746. It was known as Hart's Upper Weir at that time, no doubt operated by the Hart weir-keeping family. In the fourteenth century there was a watermill, and *Jackeswere* survived at this spot into the nineteenth century. John Rocque's survey of 1762 shows Farmer's Weir, locating it at (SU241981) near a tiny brook entering the Thames from Buscot Park. There is no trace of the weir today, as its removal was requested in 1789, though a withy-bed and island are shown at the same spot (1839 Enclosure map) on land of Mr Pryse Esq. of Buscot House.

> *The isolated village of Eaton Hastings just south of the river was once Water Eaton. Ferry Cottage near the church hints of a crossing over the adjacent river. The Domesday entry for Eaton records two fisheries and about 150 acres of meadows previously held by Gyrth, King Harold's brother. The well-established pre-Conquest hamlet had belonged to Edward the*

> *Confessor but later became the property of the knightly Poyntz family. From about the mid-thirteenth century the suffix Hastings was added when William de Hastings held the manor. St Michael & All Angels' church is set on a hillock near the river; it includes some eleventh-century stonework but was largely restored late in the nineteenth century.*

In 1895 the Thames Conservancy received funds to improve the Upper Thames downstream of Lechlade and they commissioned pound locks costing some £20,000. At an isolated spot south of Grafton Common there was a flash-weir (SU270992), probably originating in the 1300s but called Day's Weir in the 1720s, and kept by a Mrs Hart between 1789 and 1821. Flash-weir keepers commonly looked after several weirs, hence the name Hart's Lower Weir also being used. In 1828 it was called New Weir after being reconstructed two years earlier. A toll of 6d (2.5p) was levied.

In 1866 the flash-weir had the name Penstone's, but in 1895 the Conservancy removed it after building Grafton pound lock (SU271992). Permission was given in 1898 for eel-traps to be set at the lock. Today's surrounding landscape would have looked very different in 1894 and again in 1947 when serious flooding brought the river level above the top of the lock. We now find it shaded by chestnut trees; the name Grafton, from the nearby hamlet, rather appropriately means 'homestead by a grove'. A track leads north into the village, but there is currently no right-of-way over the lock south of the river.

In 1866 the Thames Navigation Act restated the Conservators' authority over the Upper Thames up to Cricklade, but local riparian owners still controlled riverbanks and particularly many flash-weirs. The Navigation Act of 1788 required regulation of a depth of 3ft 6in (1m), whereas the 1751 Act had required 4ft (1.2m). In 1869, Mr Campbell of Buscot sought permission and funding from the Conservators for dredging between Buscot and Hart's Lower Weir. Nevertheless, by the late 1800s a journey from Lechlade to Oxford could be made only by rowing boat for the first 10 miles. This brought about the decision to dredge the river as far as Lechlade with a hope of revitalising it for traffic, but use of the river for access to the T&S Canal was already in decline.

Grafton is surrounded by farmland and channels draining into Little Clanfield Brook, which was strong enough to power watermills until the nineteenth century. The Domesday Survey recorded a settlement with meadows and pastures.

Meadow names suggest why the lush grassland thrives – Duck-puddle, Frogmore ('frog-mere') and Honeyham ('a sticky meadow by the river'), a name first noted in 1323. There may once have been a prehistoric settlement as suggested by areas of crop-marks revealed by aerial photography.

Mid-nineteenth-century links with the Thames occur in Clanfield place names such as Mill Lane, Marsh Lane and a field named Withy Beds, where osiers were harvested for stabilising river banks and for binding bundles of goods to be transported on the river. Today, willows on the riverside farmland are cultivated for modern usage and an area devoted to attracting wildlife of all kinds has a nature trail for interested visitors.

The river continues towards Radcot and passes a wartime pillbox with wide views to the north and south. The river makes an abrupt change of direction at Hell Turn, not far from the sixteenth-century Camden Farm. A riverside path eventually reaches Radcot near a popular caravan park.

> *Radcot was* Redcote *in 1150 and* Rodecot *by 1351, names thought to derive from 'a reed-thatched cottage'. Saxton's 1574 county map shows Rotcot. The market charter dates from 1272 – a time when many settlements began to flourish – but Radcot failed to grow because it was too near already well-founded Bampton.*

The Medieval Bridge, Radcot (SU285994), built from Taynton stone c.1220, was a packhorse and turnpike route between Burford and Wantage from 1771. Tolls ceased in 1873.

The river divides into three channels at Radcot Bridges (SU285995); the single-span bridge in regular use dates from 1787 following the Thames Commissioners survey in 1777 and their demands for improved navigation to benefit traffic using the T&S Canal then under construction. The crossing had been upgraded in 1771 for the turnpike route between Burford and Wantage. Typical turnpike bridge tolls were 4d for a horse-and-cart and 10d per score of livestock until the toll was removed in 1878 following bridge repairs the previous year. Nineteenth-century turnpike itinerants were called pikers or pikees, sometimes in a derogatory sense. In the mid-1880s it was common practice for scores of sheep to be penned beside the river where they were washed. The monks of Abingdon Abbey recorded that they conducted a *sceap-waesce* in 944 and it seems probable that monks built the earliest bridge. The sheep-wash remained in use until the 1890s.

The single-span bridge was built over an improved channel for vessels unable to pass safely through the medieval bridge, an ancient three-arch structure (SU285994) built of Cotswold stone from Taynton quarries near Burford. Two arches of Gothic ribbed stonework date from about 1220 – Henry III's time – but the central arch is younger. Some historians say that a bridge existed by the 1150s, or possibly before the Norman Conquest. King John authorised repairs to benefit wool-traders, which suggests the structure was already growing old. A prehistoric stone axe-head was discovered near the bridge during later repairs and though not proof it was a crossing place, it implies early local habitation. There was a grant of pontage in 1312 when the bridge-toll collector was an individual named William and the revenues went to the riparian owners each side of the river. Clearly this crossing place was of growing importance on a route that had existed for many centuries.

Packhorse traders used the ancient bridge and on the parapet above the centre arch is a niche in which a statuette of the Virgin Mary would have stood before the Civil War. Many years later William Morris's daughter May was instrumental in having the old bridge repaired. Her father founded the Society for the Protection of Ancient Buildings in the spring of 1877 and she continued his work. Many medieval travellers with a packhorse train would surely have halted at wayside taverns to exchange news and gossip whilst supping refreshing ale. Today's travellers at Radcot, by road or river, can visit the old Swan Inn, which was built in about 1873 replacing an earlier structure.

In the twelfth century a more northerly channel of the river, just discernable beside the newer bridge, gave access to a wharf originally used for loading Taynton quarry stone destined for buildings in Oxford and London. The quarries were used in late Anglo-Saxon times and famous even then for top quality stone. In the 1870s coal for nearby villages was unloaded at the wharf. The channel, accessible by boats from the eastern end, is crossed by Pidnell Bridge (SU286996), rebuilt in 1863 as a four-span cast-iron and timber structure. More recently it has been renovated.

> In 1141 Matilda, countess of Anjou, set up a castle near Pidnell Bridge, probably where Garrison Field is, during disputes with Stephen over England's crown, which were unresolved until 1153 when her son became Henry II. Another quarrel took place in 1387. In a battle south of the river, troops of Henry Bolingbroke beat those of Robert, 9th earl of Oxford – a close friend of Richard II. At that time Robert de Vere (recently created Duke of Ireland), unable to escape over the damaged bridge, made his horse swim across the river. During the Civil War Cromwell's troops based at Garrison Field fought Royalist cavalry in this neighbourhood in November 1644. The effects of the hostilities seem not to have lingered to the present day or influenced the tranquillity here.

During the construction of the new channel at Radcot bridges the course downstream was significantly altered. Cartographic evidence from Rocque (1764) shows Monk Mill, possibly dating from the early fifteenth century. Greenwood's 1824 map also shows the mill and Old Eye Weir, most probably located at SU288995 near the old channel used today only as moorings. A 1740s flash-weir (SU288997) near the mill had a toll of 9d per 5 tons. By the 1790s this rose to 9s (45p) for 60 tons.

On the right bank a footpath follows the Thames downstream and crosses the disused channel at Cradle Bridge (SU288996) – shown on the 1840 map. The structure has attractive latticework timber side panels. The original towpath continued on the southern bank as far as Rushey Lock. In the mid-1950s a US Air Force bomber, based at Fairford, crashed near here and burst into flames.

From Cradle Bridge (SU288995) a path crosses Thrupp Common to the site of a weir (SU296995), recorded in 1242 and also shown on the 1833 OS map. This weir was on Wadley Stream – a Thames side channel that rejoins the river further downstream. A short distance further south was the Spotted Cow (1830 map), a stone-built inn (now Spotted Cow Cottage), only 500 yards from the Wadley Stream and very likely once a popular hostelry. In 1811 Clark's 'back weir' was repaired; this was possibly the same weir (SU296995) as Mr Clark apparently ran the Spotted Cow inn. A field named The Weir Pool in the fifteenth century might also relate to the same weir. However, a map of 1794 produced by Richard Davis of Lewknor shows a Clark's Weir on a northern side channel named Burroway Brook; doubtless the Clark family were weir-keepers.

Radcot Lock (SP296002) is another of the nineteenth-century navigation improvements; the pound lock was completed in 1892. In 1868 the flash-weir was

Old Man's Bridge (SP299002) near Bampton. The footpath crossing here formerly used Old Man's Weir, a flash-weir from 1788. In 1868 the weir piles were removed and a five-arch trestle bridge was built, but replaced in 1894 when it became unsafe.

removed and replaced by a lock. The weir toll in the mid-1700s ranged from 1*d* to 5*d* per 5 tons and in 1821 quite large boats of 40 to 60 tons were charged 1*d* per ton. In the past the lock-keeper here kept a very pretty garden for which he won a prize from the Thames Conservators. Weir- and lock-keepers often ran all manner of sidelines, many of which were disapproved of by their employers because the serious work was being 'river guardians'.

Downstream from Radcot Lock was a flash-weir in 1746, sometimes called Buck's or Clarke's weir. It included a typical footbridge with hand-rails and led to a pair of thatched cottages. 'Buck' may refer to a conical wicker eel-trap mounted on a flash-weir. In 1868 the weir piles were removed and High Bridge of trestles and five arches was built, but replaced in 1894 when it became unsafe. The river passes beneath today's timber arch called Old Man's Bridge (SP299002) where a footpath once used the flash-weir called Old Man's Weir from 1788. In the neighbourhood is a field, Old Man's Piece (1830s map), where the 'old man' was probably Mr Harper, whose name was linked with the crossing until 1884. Since at least 1866, a footpath went to the Spotted Cow inn at the edge of Thrupp Common, mentioned above. Aerial photography suggests these fields may have supported very early habitation.

South of Radcot the market town of Faringdon (prefixed 'Chipping' during the fourteenth century) has numerous seventeenth-century buildings. The settlement began with the Saxons who had a mill and a fishery and it became a 'royal manor' of King Alfred who granted the borough its market charter, but its position on a wool-traders' route in the thirteenth century brought it prosperity. Market day was Tuesday and there were monthly cattle and sheep markets and an annual Horse Fair held in February. Robert, earl of Gloucester, built a castle here supporting King Stephen against Matilda's challenge to the throne. The castle was later destroyed.

All Saints' church has a thirteenth-century south door and some interesting memorial brasses. Among them is one for John Sadler who died in 1505 and was vicar of Inglesham, near Lechlade. There is also a somewhat mutilated brass to Thomas Faryndon (1396) in armour beside his wife Margaret with their daughter Kath. During the Civil War, when the town was under siege, Parliamentarian troops toppled the church steeple with cannon-fire. In the grounds of Faringdon House (built in 1780), north of the town, there is an elegant tower on a fir tree-clad knoll where the twelfth-century castle once stood. The 100ft (33m) brick tower was built in 1935 by the landowner, Lord Berners, to a design by his friend the 7th Duke of Wellington

> *who was an architect. It is said to be the last folly built in England (that is if we disregard the Millennium Dome, now the O-2).*

From Radcot to Newbridge the countryside was prone to flooding until the common land was enclosed after the 1848 Act and the land drainage Act of 1866. The river meanders through the rather flat but lush countryside; it maybe seeks a new course or perhaps an old one. Overhanging blooms hum with bees, summer skies, full of tower clouds, fade to paler blue above the skyline hills. Dragonflies dart low over the water and forage in reeds topped with feathery brown heads. A squeaky call comes from a kingfisher flashing past and his brilliance competes with the sun sparkling on the river.

Rushey flash-weir, Buckland, 1878 (SP323001), where the Thames Commissioners operated a ferry for towing horses when changing to the left bank. This was one of the last old-style paddle-and-rymer weirs. (Reproduced with permission of English Heritage, NMR)

Where Burroway Brook flows within 50 yards of the Thames there used to be Old Nan's Flash-Weir (SP314004) and a weir-keeper's cottage (1830 OS 1in map). In 1771 the weir toll was 2½d per 5 tons, but by 1791 there were complaints that the weir needed repair as it could sustain only 10in of water. A pound lock was planned, but never built. By 1802 the weir was 'in pieces', though a toll was still being levied in 1821. By 1861 there was no trace left and in 1868 the river was widened.

Some historians say there was a prehistoric ford hereabouts. There were fords on Charney Brook (SU306007) and formerly at Burroway Bridge (SU309006). Both brooks are Thames side-streams. New Weir Ham, a field name near Charney Ford, suggests a newly built weir. Wall's Weir (SP319008) on the 1830 OS map was on Charney Brook. In the mid-thirteenth century *Lyndwere* (from linden trees?) was recorded, a possible forerunner of one of these weirs. The same map shows a barrow (Iron Age tumulus) on a knoll between Clanfield and the Thames; perhaps the 'barrow way' led to a ford.

At Rushey Lock (SP323001) the old towpath returned to the left bank as far as Duxford ford and in 1816 the Thames Commissioners operated a ferry at the lock for towing-horses. A right-of-way over the weir suggests this has long been a regular crossing place. A path from Bampton (via Fisher's Bridge) crosses the Thames to Littleworth or to Buckland via Carswell Marsh; all these settlements had pre-Conquest mills and fisheries.

The lock dates from the improvements of 1790, when the keeper was Mr Rudge – possibly the same who was also keeper at Tadpole Weir further downstream – and by 1798 Joseph Winter was weir-keeper. By 1866 the Conservancy was responsible for appointing the keepers. The stone-built pound lock had become very rundown by 1857 and in 1874 the weir was described as dilapidated. In 1875 it was repaired, then re-built in 1896 after the severe floods of two years earlier. When Henry Taunt, an Oxford photographer, travelled the river in the late 1800s he described the weir as being 'old and broken'. Today, the weir-channel operates with a modern paddle-and-rymer weir.

Rushey (meaning 'rushes island') was first mentioned in 1069. In the late Anglo-Saxon period the settlement of Chimney included a 'rush island and Berhtwulf's island'. The settlement boundary was partly along the Thames and along other streams nearby. By 1542 the bishop of Oxford was the landowner. In the eighteenth century the area was Common Land and largely undrained, but the Great Brook, probably a 'canalised' side-stream, flows parallel with the Thames from just upstream of Rushey Lock to a point near the hamlet of Shifford. It

Tadpole Bridge (SP335004) was on a turnpike route 1777–1875 and was the site of a former ferry adjacent to the Trout Inn. A plaque on the bridge marks the flood level of 1894.

drains farmland brought into cultivation by land enclosure in the 1840s and under drainage regulations of the mid-1860s, when small tributaries were dredged and fords removed.

According to the 1830 OS 1in map, Winney Wegg Wear (SP332008) on the Isle of Wight Brook (a Thames side-stream) had a toll of 4d per 5 tons in 1764 and in 1789. In 1821 the weir owner, Mr Southby, levied tolls of 3½d per 5 tons. This confirms a channel used for navigation. The Isle of Wight Brook joins Great Brook at Isle of Wight Bridge (SP333002). On Ham Lane, between Aston and Tadpole Bridge, there was Quacken Ham Ford (SP345013), which confirms local needs for river crossings.

We are now halfway between Cricklade and Oxford. From Rushey Lock a track (the old towpath) follows the northern bank to Tadpole Bridge (SP335004), which dates from about 1784 and was built for the Witney to Wantage turnpike. A 'newer bridge' was contemplated in 1796 and was probably completed by 1802. Moule's 1830 county map shows it. In 1875 the turnpike status of the bridge and toll-collection ceased by Act of Parliament. Tadpole Bridge would have been a useful crossing when many turnpike roads were being improved. Floods in November of 1894 were apparently quite severe; a plaque marks the flood level on the stonework.

Tadpoll House, south of the river (1830 OS 1in map), and Tadpole Meadow shown on Rocque's 1761 map are the likely origins for the name of the bridge and may derive from 'toad-pool'. Miscellaneous archaeological finds near the bridge suggest the Romans came here. The Domesday entry for Bampton records Salt Rights at Droitwich. The salt was brought along a network of salt ways over the Cotswolds and it was sold at Shellingford, south of the river, which suggests that salt-traders used this crossing place or Radcot Bridge in earlier times. Early medieval traders used either a ford or later a ferry. Modern travellers are pleased to find a bridge and the Trout Inn where the toll-gate once stood and where a coal wharf operated from about 1800 until the first decade of the twentieth century.

A little downstream there was Tadpole Weir (SP336004) (1830 OS map), called Kent's Weir from 1794 through to 1861, most probably referring to John Kent, a relative of William Kent who kept the Trout Inn in the late eighteenth century. The weir, also known as Rudge's between 1746 and 1827, survived until 1869 when it was dismantled. Names of weirs can be very confusing due to the practice of adding the keeper's name – often a member of a large weir-keeping family. Weirs often took the most recent keeper's name. Many private weirs were removed to improve navigation.

North of the river is the market-town of Bampton recorded in 1070 as Bemtun *– thought to mean 'a settlement near hornbeams', trees that produce a very hard wood much favoured for mill cog-wheels. Bampton grew as the 'hub' of a huge Anglo-Saxon royal estate, granted in the 950s when there was religious revival after the Viking incursions. The settlement had market-town status well before the Conquest and the Domesday Survey shows a forester called Bondi cared for the woodlands and that there were mills and fisheries, though only some would have been on the Thames as there are many rivulets in the vicinity.*

William of Normandy gave land to high-ranking churchmen and the bishop of Exeter held Bampton estate in 1203. By the mid-twelfth century the town had become very wealthy despite a rural location, but some of the older buildings of stone and thatch, a legacy of earlier times, show their humble origins with timber lintels over the doorways and windows. Until the 1780s this area was known as 'The Bush' – a wild scrubland with few roads and often frequented by footpads who were a menace to travellers.

The town's proximity to the river would have benefited the eighteenth-century maltsters importing fuel for their barley-roasting kilns, making malt for dozens of local breweries. Many fit, sturdy young men, particularly during the agricultural depression of the late nineteenth century, found work in the maltings after the bustle of harvest was over. Another important nineteenth-century industry was dressing skins for all kinds of leather goods made by women and children. The town's livestock markets and annual horse-fairs, held until the twentieth century, were further reasons for cross-river communications.

The graceful spire and miniature flying buttresses of the thirteenth-century church of St Mary the Virgin, which incorporates traces of a former Anglo-Saxon minster, can be seen from the riverside marshes. Within the church are some memorial brasses; the oldest is a half-effigy to Thomas Plimmiswood (1420), a vicar, and one (of about 1500) for Robert Halcot MA, shown wearing a choir cope. This quiet town of mostly eighteenth-century Cotswold-stone buildings has an almost unbroken tradition of Morris Dancing – a type of team dancing usually performed by men. Long ago it was customary for the men to blacken their faces to perform folk-dramas about Turks or Moors, from which the modern name Morris is said to derive. In the Cotswolds, no faces are blackened but white handkerchiefs and clothes decorated with ribbons and bells are still worn. The Yuletide Mummers Plays allude to the tussle between good and evil and are traditionally performed around the town.

South of Tadpole Bridge is Buckland, another place of Saxon origin that once belonged to Edward the Confessor. In 957 King Eadwig granted about ten hides of land to Aelfheah, governor of Hampshire, but Abingdon Abbey also owned some of the land. The Domesday entry for Buckland is Bocheland, meaning 'land granted by charter' (or land-book), making it heritable property. The entry lists a watermill, which existed until 1317, and four fisheries on the Thames, 220 acres of pastures, and a dairy, which paid the king about ten hundredweight of cheeses.

The church of St Mary the Virgin has a noticeably wide twelfth-century nave, a Jacobean pulpit and box-pews. In 1544 the manor was sold to John Yate whose mural brass of 1578 shows he lived for sixty-six years. It also shows his wife Mary and their many children. More than a century later, the estate of another John Yate passed by marriage to the Throckmorton family. The imposing Palladian-style Buckland House dates from 1757 and was designed by John Wood (the Younger) of Bath for the Throckmortons.

So much for village sojourns; back on the river we find it winding its lonely way between open meadows on one bank and tall whispering poplars on the other.

Tenfoot footbridge (SU354996) – a walker's view of the quite steep slope, which allows a 12ft headway for boats. The bridge replaced the 1830s flash-weir in 1867. The name 'Tenfoot' doubtless originates from the width of the flash-weir opening.

As the great arch of Tadpole Bridge recedes behind us there is the silvery flash as a fish flicks lazily and plops into the quietly flowing water. As evening mists form over the river a robin's clear notes may be heard. Along the way we pass some of the occasional riverside Second World War defences – derelict pillboxes – on the northern bank.

A mid-tenth-century land charter for Buckland refers to 'a weir at a nook of the Thames'. Maybe this was the same as Thames Weir (SP348002) that Griffiths noted in 1746. Thames Weir gave access to 'the old river' and to Chimney, which had a valuable fishery belonging to Bampton minster. In 1771 a toll of 1d per 5 tons was levied at the weir, which by 1821 was in poor repair. However, on a map of 1813 relating to use of the Wilts & Berks Canal the 'old river' is shown as a 'by-pass' of Tenfoot Weir and of Duxford Ferry and Weir. In 1803 there was a plan to 'canalise' the Thames from Tadpole to Shifford, but ultimately Shifford Lock Cut was built as a compromise.

The next crossing is the wooden Tenfoot Bridge (SU354996). This replaced a flash-weir of the same name, which one can assume had a 10ft-wide passage for boats. This width was also commonly used for service lanes behind many terraced properties. A pound-lock was recommended to no avail. The weir-keeper had fishing rights in 1802 and ran a public house in his thatched cottage. In 1867 pedestrians claimed the crossing was a long-held right-of-way, which the Conservators denied. Locals continually complained that the crossing was unsafe, then in 1869 Tenfoot Weir was removed and the footbridge built, though the weir's remains were evident until the early twentieth century. A crossing place has probably existed here for many centuries, as suggested by the discovery of an almost complete late Anglo-Saxon (Viking) sword in the river. In 1811 a grant was made to John Kent for towpath repairs.

This stretch of the Thames to Newbridge is through isolated country but several crossings apparently existed in the past. In 1775 there was 'Newton Weir above Duxford', and in 1793 a river inspector requested the owner (a Throckmorton) to have it repaired so that it held sufficient head of water. In the late nineteenth century there were still numerous flash-weirs and some were hard to open due to their decay. The result was that it often took vessels up to five days to travel between Lechlade and Oxford. In 1883 the T&S Canal Company complained to the Thames Conservators that there were shoals near Tenfoot weir where a rowing boat drawing only twelve inches ran aground. There were further instances of vessels running aground but exceptional droughts were blamed.

Since the construction of Shifford Lock Cut in 1898 a southern loop of the river has become un-navigable. In 1746 Duxford Weir on that loop (SP364007) was a flash-weir, later run by a Mrs Kent in 1802. It was reported to be in poor condition in 1793 and the owner was requested by the T&S Canal Company to have it repaired, but by 1869 it was removed entirely. The Domesday Survey shows Duxford with a fishery worth 25s 2d; and a mill, which survived at least until 1219.

At Duxford ford (SP370001) (from *Dudoc's*, a personal name recorded in Domesday Book) two strategically placed posts mark the optimum route for any vehicle using the crossing, and it can be crossed on foot with care and a pair of good boots. Pausing in the ford the current tugs at one's ankles and the mind turns to the river's journey through picturesque and important places before it reaches the sea. With each footfall, minnows scatter from every stone crevice. On a recent voyage upstream (navigable for small craft only as far as the ford) my arrival coincided with that of a herd of swans coming downstream. They waddled onto the stone-bedded

Ford with Swans at Duxford (SP37001). With the construction of Shifford Lock Cut in the 1890s a southern loop of the river became un-navigable. Duxford Weir on that loop was a flash-weir in 1746, and run by Mrs Kent in 1802. By 1869 it was removed.

crossing and paused to preen their feathers, and then calmly sailed on with the flow – leaving me with an enduring memory of 'swans-ford'.

When vessels plied this stretch of the old river, the ford was a crossing place for the towpath. From about 1827 until 1914 there was a 6d-toll ferry and many travellers would have crossed here between the market towns of Hinton Waldrist and Aston via the tiny hamlet of Chimney. The track from Duxford to Chimney passes a Second World War pillbox standing near the ford and then reaches Shifford Lock Cut, which it crosses on a 1990s timber bridge (SP367008) suitable for farm traffic use. Willows, poplars and ash trees make the cut pleasantly shaded. But in an adjacent meadow some twenty frisky bovines approached at a canter and skidded to a halt. Their docile eyes fell on me and I felt their warm breath in my face. They then calmly ambled off, flicking their tails.

Shifford Lock (SP371011), the last to be constructed on the Upper Thames, has the greatest fall (7ft 4in or 2.2m) of all the locks between Lechlade and Oxford, despite being in such flat farmland. The lock-keeper's garden has numerous fruit trees, possibly planted by the first lock-keeper whose cottage bears the date 1897.

> *Chimney was called* Ceomma's Island *in 1069, but began as a Saxon dependency of Bampton before 850. It became independent by the 950s and there was a Christian cemetery here. Chimney 'island' is criss-crossed by streams and ditches – doubtless cut centuries ago to assist drainage – and ideal for today's nature reserve. There are frequent references in early land charters to ditches, when used as estate boundaries. The Bishop of Lincoln owned the land, but it was in the care of Columban, first abbot of the revived monastery at Eynsham.*
>
> *Aston was known as Easton in 984 identifying it as 'the settlement east of Bampton'. In the eleventh century (until forfeiting his estates in 1082) Bishop Odo owned the common land and meadows, which were managed by 'grass stewards', who would have harvested the hay. Long before modern land-drainage schemes, much of this land would have been quite wet, except in summer when it could be used for grazing. Monastic communities owned much of the land, and estate stewards usually travelled by boat. Today, Aston is a village of stone-built thatched cottages surrounded by arable land and marshes.*
>
> *Domesday Book records Hinton (on a southern ridge above riverside meadows) with a mill and two fisheries owned by Bishop Odo of Winchester. St Margaret's church dates from the early part of the thirteenth century when Hinton became the home of Henry de Valery, who sometimes used the suffix de Sancto Walerico. His family originated from St Valery on the Somme estuary. They set up the first Hinton markets held on Wednesdays. The origin of Hinton's suffix 'Waldrist' possibly came from St Walaric who lived in the Auvergne among seventh-century peasants but ceased tending sheep to become a monk and later an abbot.*

Shifford Upper flash-weir – or Shifford Weir on the Richard Davis 1790s map (SP372014) was in such poor repair by 1876 that passage was almost impossible, but after the construction of Shifford Lock Cut, it would have become redundant or dismantled. Shifford Lower weir (SP381016?) was in use for many centuries, but maybe this was the Shifford weir reported in 1791 as 'beyond repair'. Near Shifford Farm there are two river islands – shown on the 1830 OS map – and weirs could have been located at one of these. They possibly originated as fish-weirs belonging to Eynsham Abbey, which had two weirs at Shifford recorded in 1005. At that time, the weirs were described as 'one above the lade' (ford), and 'one below it'. Domesday Book records that Shifford lands, though not extensive, yielded hundreds of eels.

Regulation of the Thames water levels with locks and weirs helped to reduce incidents like one in the 1870s when a cargo of timber was aground near Shifford for two days. It was only freed with the help of a score of men with ropes and some donkeys! There were complaints a century earlier that boats were being delayed for weeks by badly regulated depth of water. There must always have been shallows hereabouts because Shifford means 'sheep-ford'. An Anglo-Saxon land charter for Shifford refers to 'Chimney on the Thames', and in 890 King Alfred chose to hold a 'parliament' here with his bishops and earls. Much of a king's time was spent journeying through his realm, but from the seventh century onwards there were also important administrative centres, such as nearby Bampton.

A boundary charter of 958 refers to a *stan bricgge* on the Thames, which might in reality have been a stone causeway. Historians suggest that the 'stone bridge' was at Shifford, but as this hamlet was recorded in 1005 as 'sheep-ford', an earlier stone bridge seems less than likely. Nowadays, Shifford is a dispersed hamlet of a few households, with a well-kept fifteenth-century chapel close to the river.

In 1490 Shifford had a 'ferry near Longworth', and Marsh Lane from Longworth could easily have been a route to the ferry point. There was a fishery (fish-weir implied) owned by the abbot of Abingdon in 811. The Domesday entry records a mill, which existed at least until 1617. In 1829 and 1840 Sanson's Ford was noted because it required ballasting, and could have been where the ferry also crossed. (Neither location is certain, but see below.)

Approximately two miles downstream from Shifford Lock there was Limbress Weir (SP389012?), but the weir had a variety of names, one of which was Townsend's Weir. In 1796 Thomas Hart rented it and in 1789 and 1802 it was called Hart's flash-weir, or Langley's with only a hut to accommodate the keeper. In 1821 the tenant was Mr Rose of Duxford who called it Langley's Weir, yet in 1861 and 1872 it was Daniel's Weir.

Removal was recommended in 1867 and completed by 1872. The weir may have allowed crossing between Langley's Lane, which crosses Standlake Common (now mostly gravel pits) and Marsh Lane to Longworth, or perhaps over the dome of Harrowdown Hill where there is also a right-of-way. In the twelfth century Harrowdown belonged to Abingdon Abbey and is mentioned in connection with a fishery, maybe at Limbre's Weir.

Longworth (which at one time included the adjacent settlement of Draycott) was once a hamlet simply called **Wyrthe**. *It became Longworth in the fifteenth century. One of its celebrated sons was Dr Fell who was baptised here in 1625 and became Bishop of Oxford for a decade from 1676. Another well-known figure associated with Longworth was*

R.D. Blackmore, who was born here in the summer of 1825. In 1869 he created the perennially popular 'romance of Exmoor' and its characters of Lorna Doone and John Ridd.

Longworth church, most probably on the site mentioned in Domesday Book, is dedicated to St Mary the Virgin, and has some historic memorial brasses. One is for Elynor Godolphyn, 'a gentylwoman' who died in 1566, and another to John Henele (1422), rector and native of Henley (-on-Thames). There is also a brass to Richard Yate (1498) and his wife Johane in shrouds; he was doubtless related to the Yate family at Buckland.

In 1115 Standlake belonged to Eynsham Abbey and was Stanlache (meaning 'stony, sluggish stream') – an ideal place for a ford, though almost certainly meaning one across the River Windrush. Standlake became a small market town, but failed due to its proximity to Eynsham. It is a straggling village with grassy verges, and the meadows around the multiple courses of the Windrush between Standlake and Gaunt House have been exploited for gravel (in some places revealing pre-Christian cemeteries), but the water-filled pits now provide water-sport opportunities and a wildfowl haven. In 1643 Charles I's troops occupied Gaunt House and guarded Newbridge. St Giles, the thirteenth-century church has a tablet (1465) in memory of John of Gaunt. But in his time he was unpopular as he was too keen to demand taxes for military campaigns.

The River Windrush leaves the Cotswolds at Witney and turns south to join the Thames at the small hamlet of Newbridge where the Rose Revived Inn displays an amusing sign depicting a rose in a tankard of beer. From the eighteenth century this was Fair House, an alehouse with an adjacent wharf used for the discharge of Midlands coal and other goods brought up-river from the Oxford Canal. The inn's charter, first granted to Sir Edmund Warcupp, permitted fairs in March and September. The inn was largely dependent on river traffic for trade.

The name New Bridge is misleading as the six-arch stone bridge (SP403014) is said to date from around 1250. Its Taynton Stone was no doubt brought by barge along the River Windrush. It was 'new' compared to Radcot Bridge, and very probably its white limestone gleamed before mellowing to today's colour. The jutting pedestrian refuges on the bridge are supported by stout cutwaters, and the pointed arches, attractively mirrored in the river, have ancient stone ribs visible when boating beneath them.

This crossing place, halfway between Lechlade and Oxford, on a route between Abingdon and Burford, was used by medieval Cotswold wool-traders. They would have fostered nearby Witney's wool-trade. When John Leland visited Newbridge in 1535 he noted a stone causeway each side of the river that was said in 1692 to comprise twenty-eight flood relief arches on the northern approach

The medieval bridge at Newbridge (SP403014) is said to date from around 1250 and was 'new' compared to Radcot Bridge. Its Taynton Quarry limestone was no doubt brought by barge along the River Windrush, which flows into the Thames nearby.

and seventeen on the south. County maps by Saxton (1574) and Speed (1610) identify the bridge, but their maps show no roads. Later maps by John Norden, in William Camden's 1695 edition of *Britannia*, show roads in detail.

In about 1480 the bridge was considered in need of repair, and John Golafre of nearby Fyfield provided funds. Another John Golafre undertook a survey of the state of the Thames between London and Radcot in 1351. When St Peter's at Westminster owned the bridge in the fifteenth century a hermit, Thomas Brigges, was empowered under licence from King Edward to collect travellers' alms for bridge maintenance, and on numerous occasions through the centuries repairs were necessary.

Newbridge was an important crossing place during the Civil War and was attacked in 1644 and 1649. On one occasion 100 musketeers held the bridge, but were overpowered by a large army of cavalry and foot soldiers. In 1644 Sir William Waller and the Earl of Essex crossed the Thames at Newbridge in their attempt to trap King Charles at Oxford where he had a key station for attacks on London. However, the king escaped and later had a short-lived victory over Waller at the Battle of Cropredy Bridge.

Newbridge was widened from a cart-width during the seventeenth century and modified in 1791 when the Thames Commissioners said it obstructed laden barges negotiating its low arches. Repairs were carried out in 1801. The tollhouse became today's Maybush Inn.

section four

Newbridge to Oxford

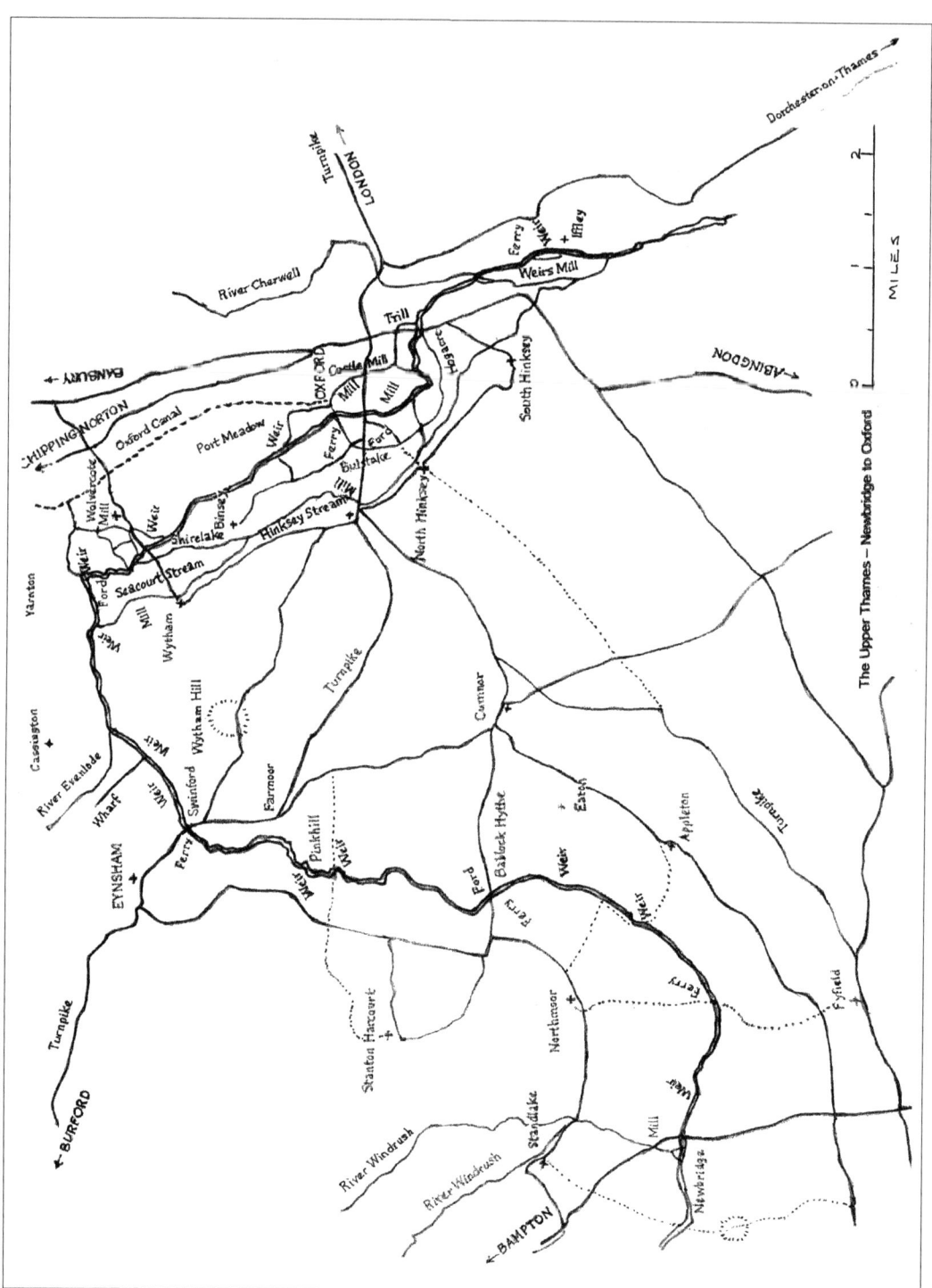

The Upper Thames – Newbridge to Oxford (14 miles).

Newbridge to Oxford

Beyond Newbridge the Thames again makes wandering progress. In 1794 it was said the river divided into 'one large and two smaller streams forming islands, one of which was inhabited'. A previous course is indicated where nowadays the river strays from the northern boundaries of Kingston Bagpuize, Southmoor and Fyfield parishes, which the river once defined. In the early 1400s there was a mill here, maybe on one of the islands beside a weir. A weir and watermill existed until 1618, possibly located on Weir Ground – a field recorded in the nineteenth-century tithe awards. The Domesday entry for Fyfield lists a fishery on the Thames worth 11s 8d (60p). Kingston's fishery was worth 5s (25p) with about 60 acres of meadows. Southmoor was called Draycott Moor in the Domesday Survey, which records a fishery worth 2s (10p) that belonged to Abingdon Abbey.

The towpath used the left bank from Newbridge to Bablock Hythe. Along the way a couple of Second World War pillboxes are reminders of the fear of invasion at that time. The valley here is broad with distant low hills most obvious to the south and the river soon begins a huge loop round Cumnor and Wytham where Wytham Hill reaches a height of about 540ft (180m). The canal engineer/surveyor Robert Whitworth proposed, around the end of the eighteenth century, a new canal to Botley, west of Oxford. It would almost halve the 24-mile journey to Abingdon, but the proposal was defeated in Parliament because tolls at Pinkhill, Godstow and Osney Locks would have been lost. From 1810 the North Wilts Canal (from Cricklade) to the Wilts & Berks Canal gave access to the Kennet & Avon Canal, and carried great quantities of Somerset coal to Oxford via Abingdon, taking much trade away from the T&S Canal and the Thames.

The next river crossing is Hart's Weir rustic footbridge (SP420010), constructed in 1879 when Hart's old flash-weir was dismantled. Rocque's 1761 map shows this crossing as Hart's Ferry, which operated until at least 1790, though by 1794 it was described simply as Hart's Weir, and in need of repair. However, it was also known as Rudge's Weir & Ferry in 1766, but by the 1890s it had become simply 'the site of Ridge's Weir'. These details suggest the ferry ceased and a flash-weir provided the only crossing with names relating to the various keepers.

Apparently, in 1789 when Mr Dalton (or Doughton) of Fyfield was owner, Mrs Grove leased the weir named Langley Weir and in 1796 the tenant was Mrs Butler.

Hart's Weir rustic footbridge (SP420010) was constructed in 1879 when Hart's old flash-weir was dismantled. Rocque's 1761 map shows this crossing as Hart's Ferry, which operated until at least 1790.

In 1802 when Mylne surveyed the river he noted 'Butler's Weir – formerly Hart's', and both names were used until 1811. However, between 1806 and 1844 OS maps show Langley Weir and in 1808 a Langley Whire was noted in connection with the proposed canal to Botley. Others called it Langley's Weir in 1872; Langley was doubtless another weir-keeper family, the name occurs at two other weirs along the Thames.

To add to all this confusion, in 1811 the weir was recorded as Jordan's Weir and a map of the same year shows Jordan's Weir Road (now Marsh Lane) from Fyfield via Marsh Farm. The name Jourdens was attached to the weir in 1826. Between 1821 and 1867 the keeper was Granny Cock and until the mid-twentieth century the keeper's thatched cottage still stood. Today, amongst tangled undergrowth, there are still the remains of the islands and a footbridge with an iron-scroll handrail similar to those on other Victorian-era Thames weirs. In 1872 removal of the weir was planned because it was dilapidated and by 1880 it had gone, being replaced by the rustic Hart's Weir Footbridge.

The right-of-way over the bridge links Fyfield with Northmoor. How often was such a remote crossing used in the past? Archaeologists have found evidence

Until the late twentieth century, Hart's Weir keeper's thatched cottage stood beside a parallel side-channel. Today amongst tangled undergrowth are the remains of the islands linked by a small Victorian-era footbridge with handrail scroll.

of a second millennium BC burial nearby and a pre-Roman lowly farmstead was built on a gravel river terrace. Iron Age finds here even suggest there was trade with northern Europe. In 1059 Edward the Confessor knew the value of Northmoor riverside meadows when he granted the living to St Denis Abbey, near Paris, and it is no coincidence that Northmoor church is dedicated to St Denis. In the late 1700s, Betty Rudge, daughter of the Bablock Hythe ferryman, married Lord Ashbrook of an Oxford College at Northmoor church. Ferrymans Farm in the village hints of other ties with the river.

> Fyfield began as an Anglo-Saxon estate of five hides (fif hidum) of land; a charter of 956 shows the abbot of Abingdon was owner. Some three years later King Edgar granted the abbey meadowland here. The Norman, Henry de Ferrers, with land at Bampton and Fyfield, seized the estate after the Conquest. In 1554 the manor was granted to St John's College, Oxford. The thirteenth-century church of St Nicholas has an effigy of Sir John Golafre who died in 1363. The church suffered a serious fire in the nineteenth century, which led to some restoration.
>
> Until the early nineteenth century most of the modern parish was moorland and arable farms, though in the 1630s oak woodlands were identified as of potential value for naval ships. Fyfield in the early 1700s was on the Bristol to Oxford coach route detailed in John Ogilby's Routes. It crossed Cothill (now Parsonage) Moor, passed Bagley Wood, to reach Oxford via Hinksey causeway. Ogilby's map shows Kingston with the suffix 'Baptist' instead of Bagpuize. The White Hart Inn at Fyfield was once a hospice dedicated to St John the Baptist. The hospice founder was Sir John Golafre of Sarsden, Oxfordshire, and during the seventeenth century it became an inn; some of the old oak roof-trusses are still visible in the hall.
>
> According to Abingdon Abbey chronicles, King Edgar gave seven hides at Kingston to his deacon in 970 and six years later King Edward the Martyr granted land to Bishop Aelfstan at a time when the estate may have included Draycott Moor. The very same year the Anglo-Saxon Chronicle recorded a great famine. After the Conquest, Henry de Ferrers let it to his tenant, Ralph de Bacquepuis. Up until the Dissolution of the Monasteries in 1538, Southmoor land was the property of Abingdon Abbey, but granted in 1574 to the Fellows of St John's College, Oxford. The name Southmoor only began around the 1740s and in the 1830s it was appended 'hamlet'.

Moving on along the river we come to the lonely Northmoor Lock (SP432021) planned in 1890 and completed by 1898, but four years earlier severe floods occurred,

The footbridge and modern paddle & rymer weir of Northmoor Lock (SP431021); planned in 1890 and completed by 1898.

as noted with a plaque set in the lock-cottage wall. The weir-channel has a modern paddle-and-rymer weir. During construction of the lock archaeological finds included Roman coins and late Anglo-Saxon (or Viking) weapons. The lock is accessible over footpaths from Northmoor village or along the left bank from Newbridge through meadows, which contrast with woodland on the opposite bank.

Beyond those woods lies the village of Appleton on a ridge above the heavy riverside soils. A land charter of 942 for Appleton tells of 'a ford on the Thames'. A *Wydewere*, recorded in 1247, was possibly at the same spot. This may have later been a fish-weir. The Domesday entry for Appleton records three fisheries; one had become about a mile long by the seventeenth century. A short distance north of Appleton is the hamlet of Eaton, its name meaning 'settlement on a river'. Domesday Survey shows Eaton was previously in the care of Haldane for the king, Edward the Confessor, and the two estates had a total of four fisheries and about 50 acres of meadowland.

Appleton has some thatched stone-built houses, a sixteenth-century moated manor house – parts of which date from King John's time – and a cosy village pub. The pre-Conquest village

Weir paddles at Northmoor Lock.

> is on land granted by King Edmund 'the Elder' to Athelstan and was known as Aeppeltun – referring to its orchards. The late Norman church, dedicated to St Laurence, has ten bells and a 200-year-old tradition of bell ringing throughout one whole day in March, which attracts hosts of visitors. A memorial brass to John Goodryngton, who died in 1518, shows a gruesome emaciated figure in a shroud. After his death his wife 'took religion in a monastery at Syon'. There is also a monument to Sir John Fettiplace (1593), whose family arrived in the fifteenth century, but died out in 1803 when there was no male heir. Edmund Fettiplace, as an Oxford–Burcot Commissioner, made a survey of the Thames up as far as Cricklade. He built Kingston Bagpuize House in 1710.

The old towpath passes the site of the eighteenth century Noah's Ark Inn, which doubtless gave its name to The Ark flash-weir (SP437033) about a quarter of a mile downstream. It was called Noah's Ark Weir in 1772; and by 1789 owned by Magdalen College, Oxford. It was mentioned again in 1811, 1827 and 1859, but was in ruins by 1866. Ark Weir Island (SP437033) is shown on the 1830 OS 1in map and Ark Ground was a field name noted in 1840. In 1832 the weir was Hart's Weir, doubtless because a William Hart had a fishery with deeds dating from 1745. The weir was where the river twists and turns and a side channel is still visible between the pollarded willows. Footbridges took the towpath over the channel and on Ark Weir Island stone foundations of a building can be traced in the grassy riverbank. Before 1866 this would have been a three- or four-room keeper's cottage. There is still a rambler rose climbing a nearby tree – an evocative relic.

The Thames continues northwards to Bablock Hythe and the Ferryman Inn, which at one time was the Chequers Inn. In 1279 it was *Babbalak*, meaning 'Babba's streamlet'. The *hythe* or 'landing-place' was used by river-traders for many years. The crossing (SP435042) may have originated before Roman times and would have linked Botley (Oxford) with more westerly places such as Bampton or Witney. A Roman stone altar, dragged from the river nearby, is now in the Ashmolean Museum at Oxford.

In the eighteenth century a handful of local maltsters worked in the barley roasting-kilns, and no doubt Bablock Hythe wharf was busy handling grain, coal and malt during the autumn and winter months. Today stone-built houses of half-timbering and thatch are set amongst extensive water meadows.

A ford was used as a reference point in a Saxon boundary clause for Cumnor lands in 904. The Domesday entry for Cumnor records two mills and fisheries worth 40s (£2), which existed until the 1530s. Tolls were given to Deerhurst Priory, Gloucestershire, which claimed ownership in the twelfth century. Thirteenth-century documents refer to 'wade-furlong' and 'ford-furlong', also to a 'ford for cattle'. One account records a causeway 1½ leagues in length (a league was about 3 miles).

Bablock Hythe Ferry probably began in the tenth century and the old towpath crossed to the right bank here. By the late seventeenth century the ferry carried carts and carriages, and Moule's 1830 county map shows the crossing place. Matthew Arnold's *Scholar Gypsy* of the 1850s crossed the stripling Thames by a ferry, which had a hauling rope. In 1894 the Thames Conservancy wanted the rope replaced by a chain and there was a long dispute over which should be used. The ferry chain was last used in 1965.

Early twentieth-century OS maps indicate a vehicle ferry; some photographs show parallel entrance slipways and ropes. Did two boats ever operate here? A typical ferryboat had a shallow draught, but maybe 25ft long and 6 or 8ft wide (approximately 8m x 2m). Livestock and small vehicles were carried and the ferry was essential to local tradesmen with their horse-drawn delivery vans. However, in the winter of 1962 it was possible to walk across the frozen river when some of the lowest temperatures occurred since the late 1880s. Nowadays a passenger ferry operates intermittently, if at all.

Cumnor was first mentioned in the seventh century when Ceadwalla, king of the West Saxons, gave land to the abbot of Abingdon. Cumnor Place was built originally as a refuge for monks in infirmity or to avoid the Plague. In the thirteenth century the monks established a grange farm on Cumnor Hil, and an annual fair was held on St Luke's Day (18 October). Cumnor House was the scene of the tragic death in 1560 of Amy Robsart, Lady Dudley, who died falling down some stairs. Her husband, Robert, earl of Leicester (implicated in his wife's death), was a favourite of Queen Elizabeth I and was lord of Cumnor manor. Lady Dudley is said to haunt the house, particularly on the stairs where she died. The tragedy provided some inspiration for Sir Walter Scott's novel **Kenilworth**.

Stanton Harcourt owes its suffix to the Harcourt family, who arrived from Normandy shortly after the Conquest. The estate, previously held by the Bishop of Bayeux, was a large one of over 200 acres with three mills and two fisheries, most probably on the Windrush tributary. Stanton is a name commonly used for a settlement related in some way to stones, which in this instance may refer to the Devil's Quoits Henge. The stone circle – mostly lost under water due to gravel digging – was south-west of the village and dated from around the third millennium BC. It was doubtless on a prehistoric route to the Berkshire Downs Ridgeway. Stanton's archaeological finds (mostly in the alluvial deposits) have been wide-ranging; some of Scandinavian origin date from the ninth or tenth centuries, but others are of prehistoric date – a bison skull, a lion jaw, a mammoth tusk and many flint tools, all suggesting a scattered population living off the land and trading across the river for many thousands of years.

Bablock Hythe vehicle ferry (SP435042) *c*.1920 was hauled across the river by rope. The crossing was an important link between Oxford and more westerly places such as Bampton or Burford. (Courtesy of www.old-england.com)

> The manor house was dismantled in 1780 leaving a battlemented tower from the 1470s structure begun by Thomas Harcourt, whose memorial brass is in St Michael's church. The family moved to Nuneham Courtenay and a new mansion was built of stone taken from the old house and transported by barge along the Thames. The family reoccupied the Stanton manor house after the Second World War and later restored it and the lovely gardens that are sometimes open to the public. In 1320 a shrine was built for St Edburga (of Bicester) who was reputedly a daughter of Penda, king of Mercia, who had become a Christian in 654 after Edburga's death four years earlier. The twelfth-century church near the manor walls has a rare thirteenth-century rood screen. The Harcourt Chapel has

> *a monument to Robert Harcourt, standard-bearer to Henry (VII) Tudor at the Battle of Bosworth Field in 1485.*

About 1½ miles downstream from Bablock Hythe, footpaths converge on the site of the former Skinner's Weir and wooden footbridge (SP438065) (1830 OS 1in map), both dismantled in 1908. The Skinner family were flash-weir keepers for many years and nearby they kept the reed-thatched stone-built Fish Inn. In the 1790s Boydell referred to Langley Weir in Stanton Harcourt; apparently Skinner's Weir was sometimes called Langley's. Surveyors noted a Langley Weir at various times between 1746 and 1827. Both names were still used in 1832 and 1861, but the Ordnance Survey favoured Langley in 1876. In 1802 Mylne recommended that a pound lock should be built to replace the flash-weir. The weir's timber footbridge was damaged in a 1930s fire caused by some boisterous university undergraduates.

Across the river, a footpath crossed Farmoor Reservoir on the central divider following a route used in the past before the reservoir's construction. In the early 1980s a new river footbridge was planned, but not yet built. On the riverbank nearby, Rocque recorded a mill in 1761, perhaps for Lower Whitley Farm or for Pinkhill Farm; however, all traces were lost by the end of the nineteenth century.

> *Close by but not visible from the river is the vast Farmoor Reservoir, which is frequented by wildfowl, anglers and dinghy sailors. It was begun in the 1970s using a tract of land known as Farmoor Common where there was a pond or 'ferny mere' in an area exploited for gravel until the reservoir was built. The Romans had a farmstead near the river, and a route along the flood-plain edge ran south-westwards through Appleton parish. Between the river and the reservoir there is a wetland conservation area begun in the 1990s where a wide variety of wildlife, birds, dragonflies and flowers can be seen.*

The Thames continues its northerly flow to Pinkhill Lock (SP441071). A flash-weir was once located near Lott's or Luck's Hole, a sharply curving stretch of water that was by-passed in 1899. In 1791 Lord Harcourt owned the weir and surrendered tolls to the Commissioners. They, by agreement, would maintain

The site of the former Skinner's flash-weir where footpaths converge on the river.

Skinner's Weir and footbridge, c.1890 (SP438065), dismantled in 1908. The Skinner family were weir-keepers for many years and they also kept the Fish Inn nearby. (Reproduced with permission of English Heritage, NMR)

and repair the weir. In 1793 Mrs Hill was weir-keeper and Mylne recorded timbers obstructing navigation. The weir toll was 4s (20p) in 1796 but there was no tollhouse, not even by 1832, and no lock-keeper's cottage until 1880. The T&S Canal Company later requested the owner to have the weir repaired. Mrs Hill was also keeper at Godstow and Osney locks. The weir, in ruins in 1872, was repaired in 1877 and a new construction planned in 1880 was completed within twenty years. In 1896 the Conservancy attempted to buy out the weir owner.

Pinkhill Lock was modernised in 1932 and in the late 1990s the lock was restructured again. River levels here are monitored using modern technology and Farmoor Reservoir maintains, by extraction or injection of water, a balanced Thames flow. This should help prevent a reoccurrence of the extensive floods of 1894 and 1947, which almost inundated the lock-keeper's pretty cottage garden. Floods can make the river 'invisible' when its course is lost, but in severe drought, the river can also be lost. However, problems were encountered yet again in the winter of 2000–01 when river levels reached all-time highs.

Farmoor Reservoir (SP438065), invisible from the Thames, has the important role of balancing river levels. Until recently a public footpath crossed the reservoir on a narrow walkway.

The restructured weir has a fish-pass for trout and maybe also for slower roach travelling upstream.

> From the thirteenth to the late eighteenth century, Pinkhill was Pincle, from Old English meaning 'a small ditched enclosure'. The Domesday Survey recorded two fisheries, but by the 1530s there was only one. Nearby is Limb Brook, a tiny tributary whose name may derive from lime-trees. It joins the Thames from the north-west after a tortuous course through the riverside meadows.

Very soon the river comes to Swinford where there is a splendid privately owned bridge (SP443086), which was opened in the summer of 1769 near the spot where John Wesley almost drowned five years earlier on his way to preach at Witney (he lived a further twenty-seven years). Shortly after its completion the bridge was tested almost to the limit by severe floods. As early as 931 a *swine-ford* was

Pinkhill Weir stream (SP441071) early 1900s. A flash-weir was formerly located near Lott's Hole, a sharply curving stretch of water that was by-passed in 1899. (Courtesy of www.old-england.com)

identified when King Athelstan granted revenue from a fish-weir to St Mary's Abbey, Abingdon, and the rents were often paid in eels. The monks of Eynsham Abbey constructed fish-weirs from stone and in 1677 their old weir was recorded as Swithin's. Documents of 1362 and 1483 show Swynford also had a wharf.

From at least 1692 a ferry carried horse-drawn vehicles, but the new bridge supplemented the ferry and satisfied growing demands for a safer river crossing. The ferry was initially run by monks of Eynsham Abbey. A 1299 record shows that ferry rights belonged to the abbey but land each side of the crossing had long been the property of St Mary's at Abingdon. There is a record of the ferryman in 1759 taking the vicar of Dean Court (on the old road from Botley, Oxford) across the river where the vicar took hold of the reeds on the Oxfordshire (western) bank – symbolically laying claim to 'the way across the water'. The ferry ceased in 1777.

After Henry VIII seized all the monasteries and their possessions in the late 1530s, the crossing had a succession of lessees. The last one, before Lord Abingdon, was John Winter in the early 1760s. The multi-arch stone-balustrade structure, designed in 1768 for the 4th Earl of Abingdon, is a scheduled Ancient Monument and has recently received some renovation. It is one of only two toll bridges over the Thames. The toll authorisation, dating from George III's time, was granted in perpetuity, originally at one penny per wheel (which later

Swinford Toll Bridge (SP443086) is privately owned. It opened in the summer of 1769. From at least 1692 a ferry carried horse-drawn vehicles, but the new bridge supplemented the ferry. The first ferry was run by monks of nearby Eynsham Abbey.

included a car's spare), and was levied at the same rate as the ferry. Tolls are still collected at a tollbooth.

> The ancient westward route from Oxford was through Botley and over Wytham Hill to Swinford Bridge, where the river crossing probably began in the Stone Age. The route was upgraded in 1751 to a turnpike road and continued in use until 1835. However, in 1810 a newer turnpike (Eynsham Road) via Farmoor was opened and by the 1820s used by mail-coaches. Thomas Moule's county map of 1830 shows Swinford Bridge, which until 1936 was on the main route between London and South Wales. Today, most traffic prefers the A40, built in the mid-1930s.
>
> The nearby market town of Eynsham has an Anglo-Saxon Chronicle entry for the year 571 when Cuthwulf, leader of the Saxon Gewissae clan, fought the Britons and captured the town from them. By 821 a Mercian king owned the settlement then known as Egene's ham. Many religious communities were founded in the early eighth century, and in 1005 a Benedictine monastery foundations were laid out over earlier buildings. It was endowed with woodland, pasture and mills to provide income and food for the monks. With the upheavals of the Norman Conquest the monks abandoned the abbey, but by 1095 it was revived with more lands at Yarnton, Shifford and St Ebbe's Island, Oxford. One monk became guardian of the estate owned by the Bishop of Lincoln.
>
> The Domesday entry for Eynsham records a mill worth 12s (60p) and that hundreds of eels were gathered from the river. In 1342 Abbey Mill was driven by Chil Brook, a tiny Thames tributary that the monks diverted to fill the monastery fishponds in the open land at the end of Abbey Street. The mill, which worked until the nineteenth century, was employed in papermaking in the eighteenth century, using the iron-free water of the stream.
>
> The Cotswold stone-built town grew around the abbey gates and soon became important. The Dissolution of the Monasteries had little effect, unlike for some other abbey towns. The market, established in Stephen's reign and held by king's charter from about 1250, was still held in the eighteenth century. Nowadays the shopping centre is round the same Market Square. Eynsham church, dating from the thirteenth century with a fifteenth-century tower and wall paintings, is dedicated to St Leonard – a favourite of the Benedictine brethren.
>
> North-west of the town, near the sparsely populated Barnard Gate on the 1750s turnpike road to Burford, was the Britannia Inn (now the Boot). This was the site of Tilgarsley, a medieval settlement, which was as large as Eynsham in the thirteenth century but was abandoned by 1350 after the Black Death had decimated the village. This was a bad time for

> *many places; the population declined in once-flourishing villages due initially to poor harvests from about 1325 and was accelerated by The Plague in the late 1340s.*

In a short distance the river reaches Eynsham Lock (SP445086), a popular walk destination from Eynsham as it passes the Talbot Inn. The lock replaced a flash-weir, one of many rebuilt as pound locks in 1928 after the Thames Conservancy had stated in 1892 that all flash-weirs must be removed. The flash-weir, which in all probability began as a fish-weir for the abbey, was called Bolde's Weir in 1791 and owned by Lord Abingdon. By 1795 it was badly decayed and not holding water. Mylne pressed for a pound lock in 1802 – without success – and by 1859 passage was said to be dangerous. The troubles might partly have been because there was no keeper until 1850. In 1886 the old paddle-and-rymer weir was replaced. More recently there have been improvements and water level at the weir is monitored scientifically. The excessive rains of winter 2000/01 put flooding statistics into the record books; many other locks experienced problems.

A path from Eynsham Lock leads to a substantial oak-beam footbridge over the reed-lined Wharf Stream, which once linked the Thames to an eighteenth-century wharf near the Talbot Inn. In the mid-1920s a sugar beet and dried crops company used railway sidings adjacent to the Wharf Stream for a period of about four years until the venture failed.

> *The Wharf Stream was bridged by the Witney & East Gloucestershire Branch Railway, which opened to passenger traffic in 1861 and closed 100 years later. An initial supporter of the line was Charles Early whose name is famously linked with the world-renowned Witney blanket business, which started as a cottage industry and in cloth-fulling mills in the 1220s, and expanded in the seventeenth century when the town weathered a period of decline. From the stone-built town originating in Anglo-Saxon times, there was an annual autumn dispatch of bulk orders of blankets by rail to London stores, and a daily dispatch of gallons of milk collected from local farms. Eynsham station was busy with thousands of passengers and tons of freight per year, much of which would have been Thames trade in earlier centuries. During the Second World War the railway benefited many of the RAF stations in the eastern Cotswolds.*

The Thames-side path continues along the western edge of the ancient 600-acre woodland on Wytham Hill, which was bequeathed to Oxford University in the 1940s for scientific research. It is a haven for wildlife and the haunt of many songbirds such as warblers and nightingales.

Opposite Wytham Wood is Cassington Cut, which gave access from the Thames to a mill and wharf beside the Cassington to Eynsham road and also to the river Evenlode when it was navigable and used by vessels transporting stone slates from Cotswold quarries. In the sixteenth century the cartographer Saxton named the river as Evenlade, which suggests the stream had shallows where a ford might exist.

The village of Cassington was Cersetone in Old English, meaning 'a farmstead where water-cress grows', doubtless true along the Evenlode, which is shallower than the Thames. It is a quiet village with a few stone-built thatched cottages. Cassington and Eynsham were closely linked, not only by a medieval port-way, but because Eynsham Abbots appointed the vicars of Cassington. From the river it is possible to glimpse the tall spire of St Peter's church, begun about 1103 by Geoffrey de Clinton. It has some interesting memorial brasses; one for Robert Cheyne, a king's esquire in about 1414, is shown in a foliated cross, and another shows Thomas Nele in a shroud. He was a professor of Hebrew at Oxford in the latter part of the sixteenth century.

It is said that in 1796 there was Clay Weir located between Eynsham and King's Locks, but its site is uncertain. However, old maps show hill-slopes south of the river named Farther Clay Hill (1726) and Hither Clay Hill (1814). Before 1974 the county boundary crossed the river at SP454092 and it could be assumed the boundary used the weir's location as an identifiable spot.

A boundary charter of 957 mentions 'Eanflaed's crossing on the Thames', which might have linked Wytham with Cassington. There were many early monastic settlements in this area and would doubtless have had communication links, perhaps with a ferry. Wytham's records of 1252 refer to *horseyelake* (horse island lake) and a Horsemere Lane appears in Cassington records of 1833. Today that lane, flanked by ash and elm trees, runs towards the river at SP462099, though today it peters out just 25 yards from it. This spot was perhaps where horses came to be washed. Perhaps Eanflaed's crossing was here, too, or maybe it was where Clay Weir was built. Who Eanflaed was and why the crossing had her name are not known.

The Thames continues past Yarnton Mead where Eynsham Abbey farmed meadows and pastures in the eleventh century, which the Domesday entry records as being larger than forty acres. At that time, Wadard, one of Bishop Odo's knights who appears in the Bayeux Tapestry, held the lands of Yarnton Mead, which included a mill, and a fishery yielding hundreds of eels.

> Eynsham Abbey's wealthy farmers kept scores of sheep until the mid-fifteenth century when disease killed most of their vast flock. Recent archaeological work has revealed a Bronze Age settlement at Yarnton and a seventh-century Saxon occupation site in Yarnton meadows where they raised cattle, sheep and pigs and traded in woollen cloth (presumably along or across the Thames). Another find was the site of a late Roman cemetery.

The river next comes to Hagley (*old-isle*) Pool where a spillway into a side stream is crossed by a towpath footbridge (SP472101). This is Wytham (or Seacourt) Stream, which in Saxon times was *Seofecan's stream*, and a significant stream well into the tenth century. From Hagley Pool, the most northerly point on the Thames through to south of Oxford, a network of streams runs parallel with today's navigation channel, but in the past they were all merely braids of one river. In 1793 the Seacourt Stream was important enough to mark the county boundary. It was called Shirelake Ditch at Binsey. Markers still stand beside this western channel, which defined the boundary between Oxfordshire and Berkshire until 1974.

The Seacourt and Botley Streams

Seventeenth-century maps by Speed, Saxton and Morden show the Thames with at least two significant parallel courses from Yarnton Mead to South Hinksey. The 1833 OS 1in map shows that the Seacourt Stream ran almost parallel with and about three quarters of a mile west of today's navigation channel. However, in the 1790s today's channel was improved and the riverbank raised 3 or 4ft (0.9–1.2m) but flooding good meadows at Yarnton. In 1866 the Thames Conservancy defined four River Thames districts between Cricklade and Oxford; the final section ended at Botley Bridge (SP491063), which crosses the Seacourt Stream, suggesting the former significance of this stream.

Long ago the many streams of the Thames were harnessed to drive mills and in about 1250 the Seacourt Stream drove Wytham Mill, property of the abbot of Abingdon until 1540 when it was granted to Sir John Williams. At that time there was also a fishery called Westman's Wynde. The 1833 OS 1in map shows a mill at a point where mill ruins exist today (SP477094). Also at that time there was Linch Farm, near Wytham, possibly with a man-made bank (or linch) for a track raised above flood levels. The alteration of river levels reducing the flow of the Seacourt Stream would have affected mills thus leading to their demise. The much-overgrown Seacourt Stream passes beneath a humped bridge (SP477088) of some antiquity near to Overford Farm, which suggests that there was a ford here before the bridge was built.

> *The neat village of Wytham grew in the mid-tenth century on land belonging to Abingdon Abbey. Its name is thought to come from Old English meaning 'village in a river-bend', probably referring the huge curve of the Thames round Wytham Hill. It is said that Offa, king of the Mercians, built a fort on the hill in the eighth century. In the medieval era, de Wythams took the name of the village previously owned by the Harcourt family, but now owned by Oxford University. There are a few stone-and-thatch dwellings and a sixteenth-century manor with an attractive dovecote. All Saints church, with a statue of Queen Elizabeth I, was rebuilt in about 1811 but the east window is, according to local tradition, said to have come from the medieval Cumnor Place, which stood some distance to the southwest until 1810 when it was demolished.*

In the thirteenth century a crossing over the Seacourt Stream linked Binsey with the lost village of Seckworth – pronounced 'secoorth' locally – and in the fourteenth century a route through Binsey to Wytham is said to have crossed the Seacourt Stream. There was a fulling mill on the Seacourt Stream from the thirteenth century and a possible location was where the 1873 county boundary crossed the Seacourt at an island (SP485078) almost opposite where the now-deserted village once stood.

> *Seckworth folk gave shelter to pilgrims, both humble and wealthy, who patronised St Margaret of Antioch's well at Binsey until it was de-sanctified in 1639 by Alderman Sayle of Oxford,*

> but the tales that as many as two dozen inns and hostelries gave pilgrims accommodation seem rather far-fetched. St Margaret was martyred by Emperor Diocletian and became the patron saint of women in childbirth. The well's healing waters were believed to cure eye troubles. Binsey church may occupy the site of an eighth-century chapel founded by St Frideswide, who is said to have prayed for the well to flow. Seckworth flourished in the ninth century and grew to a sizeable settlement with its own church from the eleventh century. It underwent a revival in the thirteenth century, but today only hummocky remains of the deserted village exist on the stream's western banks.

There is a modern bridge (SP483079) near the deserted village for the 1959 A34 (T) western bypass.

The Seacourt Stream divides at a place once known as The Dunge (SP492069) and becomes the Hinksey Stream flowing west and the Botley Stream identified on the Richard Davis 1790s map. Botley Stream very soon joins the Bulstake Stream, which we shall explore later, a little north of Bulstake Bridge on Botley Causeway. At Bulstake Bridge, the 1876 OS map names the river as the Thames.

The Hinksey (or Seacourt) Stream

The Hinksey Stream flows under Botley Causeway at Botley Bridge (SP491063), also called Seacourt Bridge. The causeway runs westwards out of Oxford to Botley – an undeveloped area until the 1940s. Nevertheless, the predecessor of the causeway was an important highway from the early thirteenth century; in the sixteenth century a local landowner gave money for its repair and rebuild and it was then raised and made a turnpike route in 1767.

A little south of Botley Bridge was a corn mill located in 'mill mead' in the twelfth century and driven by the Hinksey Stream. It was mentioned in 1637 and again in the 1720s, but in 1344 there was a complaint that Botley Mill and fishery suffered because King's Mill leat for Castle Mill at Oxford (discussed later) took too much water from the Thames. In fact, Botley Mill only went out of use in the late 1930s, thus surviving longer than Castle Mill. Early monastic records are full of similar complaints of violations of 'water-rights'. The Hinksey Stream here formerly marked the county boundary.

Hinksey is three settlements, distinguished as North, South and New Hinksey and, according to a document of 925, 'oxna-forda was by the road to Hinksey'. This agrees with the belief that a prehistoric track forded the Thames near Hinksey and then crossed the Berkshire Downs. A Celtic dagger was found in the Hinksey Stream and such finds are believed to be evidence of ancient crossing places. Historians calculate that a Roman route crossed the multitude of Thames braids at Hinksey before heading towards Wantage. From North Hinksey village a footpath runs south-west almost straight for 3 miles towards Bessels Leigh where it joins a sixteenth-century coach route to Faringdon that was turnpiked in 1768. It is also possible that an ancient route from Hinksey went through Cumnor to Bablock Hythe ferry for travellers heading westwards for Witney and Burford.

Hinksey was Hengestesige *(Hengist's island) when King Caedwalla made a gift of the land in the seventh century to the abbey at Abingdon. In a land charter dated 957 King Eadwig granted some 20 hides of land at Hinksey, Wytham and Seacourt to the Abbey. Anthony Headley was the miller and records show the mill was burnt down in 1656. St Laurence's church is largely unspoilt despite some restoration. It includes a memorial to the Fynmore family. One of Hinksey's famous inhabitants was John Ruskin who rented a cottage near the Fishes Inn.*

Hinksey Ferry, on the 1876 OS map, was near the Fishes Inn. Behind the inn a footbridge (SP496055) leads to a field path that at one time crossed Osney Mead and along Ferry Hinksey Road to Botley Causeway. The ferry was mentioned in 1467 when Botley Causeway was constructed and a path to the ferry was also made. Earlier mention of the ferry occurred in 1370. In 1539 William Bulcombe and his wife worked a ferry, ownership of which in 1658 passed from the Fynmore family to Brasenose College, Oxford. When the ferry ceased, this route was diverted to a track called Old North Hinksey Lane – now used as a cycle-track – with a stone cobble surface. At its western end this track crosses the Hinksey Stream on a stone-built bridge (SP495056) (1876 OS map). But it might have older foundations; there are hints of a ford in the shallow water below. A county boundary stone stands nearby.

A modern rural track from North Hinksey to South Hinksey crosses the Hinksey Stream on an unremarkable iron-railing bridge (SP506051). A track between South Hinksey and New Hinksey crosses the Hinksey Stream at the curiously named *The Devil's Backbone* (SP512045) shown on the 1830 OS map.

The stone bridge (SP496055) built in 1901 on Old North Hinksey Lane across the Hinksey Stream where there may be remains of a ford. The lane has a stone cobble road surface and was on the old route from Oxford to Faringdon. Hinksey Ferry operated near the Fishes Inn in North Hinksey Village. When the ferry ceased the route was diverted to Old North Hinksey Lane.

At South Hinksey there was once a ferry, possibly at SP518038 across the Hinksey Stream; in all probability near where Abingdon Road dual-carriageway is now, and where a mill existed in the 1800s. An Anglo-Saxon boundary charter refers to a 'stone ford', and to a (westbound) '*higweg*' (highway). John Ogilby's *Routes of a Hundred Roads* gives details of this route to Bath, which passed Bagley Wood with 'furse and fern on both sides'.

The Hinksey Stream then joins the Weirs Mill Stream near Oxford's southern by-pass (A423).

Weirs Mill Stream

Weirs Mill Stream flows parallel with the Thames from SP521047. Since the mid-1100s the Weirs Mill worked 'at the langford'. There are numerous modern water flow controls along the stream. It is crossed by a footbridge at SP522042, and on Weirs Lane (B4495) by Donnington Bridge (SP522043), which dates from 1926. The bridge replaced a ferry, shown on nineteenth-century maps, superseding the ford. The stream continues beneath the southern by-pass (SP521034) and re-joins the Thames at SP525028.

Returning to the Thames mainstream, which we left at Hagley Pool to explore the side streams, we continue beside Yarnton Mead and Oxhay Mead (1770s). The river divides either for entry to King's Lock, or into the Wolvercote Stream and onwards to the narrow Duke's Cut and into the Oxford Canal. The Cut, privately built by the 4th Duke of Marlborough who owned the land, was named in his honour, and constructed in 1789. Duke's Cut Lock is fed by water from the Thames, but in the past the canal water level was often above that of the river, except in times of flood. In those days the lock gates could open either way. Alteration of the level of the Thames in the 1790s changed this situation.

The Oxford Canal was one of Britain's earliest for which Royal Assent was given in 1769 to the engineer James Brindley on a salary of £200. Sadly, Brindley died in 1772 and so was unable to see the project to its completion on New Year's Day 1790, celebrated with military band music and boats making their way to the terminus wharf near Hythe Bridge at Oxford. However, there were inherent problems with Brindley's design using narrow locks of 7ft (2.1m), which slowed progress of paired boats. But the canal became an important link carrying Midlands coal via Isis Lock to the Thames and London. Isis Lock with its single gates and elegant iron footbridge is a rural haven despite its proximity to the city. The Canal passed Jericho wharf and a nineteenth-century iron foundry that produced a wide range of decorative and functional items. The Oxford Canal was superseded when the Grand Junction (now Grand Union) Canal, providing a much shorter journey between the Midlands and London, was opened in 1805. The New Road Wharf at Oxford was filled in and became the site of Nuffield College in 1937, but as a reminder of the canal era a monument in the form of a large capstan is located near Hythe Bridge on the site of the former wharf and terminus.

Wolvercote Stream

On the bank of Wolvercote Stream at SP485105 is an early twentieth-century 'County Highways'-style signpost indicating the way to Duke's Cut and to the Thames.

Wolvercote probably began as a Roman settlement and at Domesday it was a rural village with large areas of pasture and meadows. In the 1540s Wolvercote manor estate was granted to George Owen and a weir was mentioned then. The 1950s A34 trunk road (SP487101) crosses the Wolvercote Stream, which drove the Duke of Marlborough's corn mill (SP487099) that was converted during the Civil War to forge swords. In the seventeenth century the mill became a paper mill for the University Press. The 1834 Enclosure map shows Millway Furlong.

The stream flows to Lower Wolvercote Bridge (SP486094) where road tolls were paid until the mid-twentieth century. On the bridge is a large pink-and-black granite plaque, which commemorates two Royal Flying Corps

On the bank (SP485104) of Wolvercote Stream is a 'County Highways' style signpost indicating the ways to Duke's Cut, the Oxford Canal, and to the Thames.

At the old King's Weir (SP479103) a boat-slide was built in 1888 for smaller craft so that the weir could maintain the river level. This old boat-slide is at Iffley Lock, south of Oxford.

airmen who died nearby in 1912 when Port Meadow was used for aircraft landings.

On the Thames, King's Weir (SP480104) was identified as *kyngiswere* in 1182 and specifically as a fish-weir in the 1450s. A 1405 document refers to *Cayser's Were* (doubtless from *cāsere* in Old English) and presumably another name for King's Weir. By 1679 King's *Weare* allowed regular river navigation, and the weir evolved as a combined flash-weir and flash-lock with a single pair of timber gates. Early in the nineteenth century King's Weir was ruinous although some repairs were carried out. A contemporary drawing shows a structure of badly rotting timbers. In 1817 a modern pound lock was recommended, but not built until later. The lock-keeper's cottage was built close by, but in 1891 the keeper lived on a houseboat. By 1888 a boat-slide had been added beside this lock.

In 1927 a more westerly channel was cut for a new King's Lock (SP479103) with one of the shallowest falls on the Thames. A new keeper's cottage was built at the same time. Lord Desborough of Taplow (of the Board of Conservators) officiated at the opening in 1928. The pound lock suffered devastating floods in 1947, a recurring river problem and not always in winter. In the middle of the thirteenth century summer rainfall locally caused floods that swept away trees, houses, mills and bridges.

The Thames turns south after King's Lock and makes sweeping bends past Pixey Mead – farmed by Godstow Abbey in the twelfth century and mown to provide hay for oxen, sheep and draught horses. Pixey Mead is famous today for a profusion of wild flowers. When Godstow Lock was built in the 1790s there were complaints that because the river was raised 3 or 4ft (0.9 to 1.2m) it destroyed the 'summer ford' at Pixey (ford location unknown). The wide expanse of the river today may differ greatly from the late sixteenth century when weeds and flags were said to 'choke the shallow waterway'.

The next river crossing is the modern concrete Thames Bridge (SP482094) constructed late in the 1950s for the busy dual carriageway of the A34 (T). The lapping sounds of my dinghy riding the waters was drowned by the intrusive traffic noise so audible on the river where there would otherwise be quiet moorings beside the sedge and reed mace edging the meadows.

In a short distance the river divides for Godstow Bridges where an old route westwards from Wolvercote crosses the Thames. A 1676 county map by Robert Morden clearly shows this as a major route via Swinford Ferry to Burford. Sir John Walter sold Godstow Bridge and Weir (SP484093) to the Duke of Marlborough who immediately set about repairing the old stone bridge structure, which originates from about 1278, but has been rebuilt many times. This older Godstow Bridge has one pointed arch and one rounded, perhaps to allow more headway for trading vessels.

The weir doubtless existed in 1180 when a *werehamm* was recorded and may have created shallows for a ford. Perhaps it later became the site of a flash-weir. By 1788 the Thames Commissioners considered the older bridge arches too narrow for boats. A new navigation channel was cut and another twin-arched bridge of brick and stone was built (SP484092). Beside the river, which teems with fish, is the Trout Inn, begun in the 1130s as the hospice of a nunnery on land west of the river. From the terrace of the inn a traditional timber footbridge (SP483095) crosses to a tree-clad island between the bridges.

Godstow's name derives from stow *– a place of religious assembly. In 1134, Lady Edith Launcelene founded a Benedictine nunnery for gentlewomen funded by their families and she became the first abbess. The nunnery was consecrated in the presence of King Stephen in 1138,*

> but was surrendered to Henry VIII in 1539 and became a private house. Henry II's mistress Jane Clifford, 'the Fair Rosamund', was educated at Godstow and when she died in 1175 in suspicious circumstances she was buried here. Her tomb soon became a shrine, but Bishop Hugh of Lincoln was sickened by the reverence given to a concubine's tomb and ordered its removal in 1191. He also tried to ban all worship of river nymphs and sprites. Henry VIII gave the nunnery premises to Dr George Owen whose descendants were Royalists – one of whom supported Charles I during the Civil War. The premises were in private hands after the Reformation and were badly damaged during the Civil War by Cromwell's commander, Sir Thomas Fairfax. The ruined chapel walls, once ivy-clad, are allegedly haunted; a recent sighting was in the 1970s.

Downstream from the Trout Inn the Thames soon reaches Godstow Lock (SP485089) completed by 1791 under the navigation improvements required by the 1771 Act. The chosen site was known as Woodward's Hole, though an earlier plan was to build further downstream at Black Jack's Hole. The first lock-keeper was a Mrs Hill, who also kept Pinkhill and Osney Locks, but by

Godstow Bridges and Trout Inn (SP484092). In 1788 the Thames Commissioners considered the 'old' Godstow Bridge arches too narrow for boats. A new navigation channel was cut and a second twin-arch bridge was built. (Courtesy of www.old-england.com)

1798 James Bishop was lock-keeper and received seventeen shillings per month (85p). Although Godstow Lock had a keeper in 1798, his successor lived on a houseboat nearby in 1890 because a lock cottage was not provided until 1896. The weir was rebuilt sometime in the nineteenth century. The lock was deemed dangerous in the 1890s but only rebuilt in 1928. Lord Desborough presided at the reopening ceremony. The lock gates have since been converted to electric operation and improvements made to the weir-channel in the late twentieth century. Thames Conservancy badges are prominently displayed on the lock cottage wall.

After Godstow Lock, the river becomes noticeably wider alongside the vast expanse of Port Meadow and Wolvercote Common. Over the centuries the Thames has carved many courses through the Oxford clays and sands, and in dry weather some of them show up as crop-marks in the grasses. In the historic past,

The Medieval Bridge at Godstow (SP484092). The river divides at Godstow where an old route westwards from Wolvercote crosses the Thames. A 1676 county map by Robert Morden clearly shows this was a major route via Swinford Ferry to Burford and the west.

the land was deliberately flooded to attract wildfowl for shooting. Despite later land drainage schemes Port Meadow is still prone to flooding at certain times of the year.

A little upstream of Black Jack's Hole – a rather marshy area – a river overspill confirms that the river's main course has been raised. This was a stream in the 1790s shown on the Richard Davis map. The overflow passes Binsey church and through land shown as 'liable to flood' to reach the Seacourt Stream. Opposite Port Meadow, beyond the built-up riverbanks, some fields appear to show signs of medieval ploughing ridge-and-furrow, but now laid down to grass.

> *From Godstow we become aware of the distant towers and spires of Oxford – a city shimmering in the haze of summer. Wolvercote Common has been continuously used for grazing stock since Anglo-Saxon times and was farmed by Godstow Nunnery. Domesday Book describes the area as 'pasture outside the walls (of the city)'. William of Normandy granted the 350-acre Meadow to the Merchants Guild of Oxford. Today, the unimproved grassland has many rare plants and is grazed by cattle, ponies and horses. It is also frequented by wildfowl such as Grey Lag and Canada Geese, which often arrive en masse with much calling of their evocative cries.*
>
> *Almost in the middle of Port Meadow is Round Hill, which is a Bronze Age tumulus, and archaeologists have shown that Iron Age farmers inhabited the area. Land drainage legislation in the eighteenth century meant artificial pools were built for wildfowl shooting, but in the late 1800s shooting became prohibited. Nowadays, nearby reservoirs provide a protected environment for breeding birds. Early OS maps show that the Meadow had a race course, but nowadays horses are free to roam and graze.*

West of the riverside fields is Binsey, *Byni's island* in 1122, stranded on land between the Thames and the Seacourt Stream. Archaeology has revealed Anglo-Saxon occupation here in the eighth century, as doubtless on other Thames 'island' sites from the sixth century. Recorded in 1352, Binsey Ford (location unknown) crossed the Thames and was used until the nineteenth century when barge towing-horses crossed to the western towpath as there were disputes in 1798 about the left bank towpath. Was Binsey Ford a crossing on an ancient route from Oxford to *Eanflaed's* crossing near Cassington?

Wildlife at Port Meadow.

> *The thatched Perch Inn at Binsey, barely 50 yards from the Thames, dates from the seventeenth century but was reconstructed in the 1970s after a destructive fire. It has an attractive willow-shaded garden. Binsey's stone-built church, 700 yards to the north-west, probably dates from the twelfth century. The first vicar of Binsey became Pope Adrian IV. In the churchyard there is a sacred well dedicated to St Margaret of Antioch, who died in 304.*

The river divides at Medley Weir (SP498073), which used to control the flow to Castle Mill Stream. Weir Cottage stands by a small weir as a reminder. The weir-keeper's cottage was built in 1839, and a Bossom was weir-keeper in 1854. After the Floods Inquiry in 1883 the Thames Conservancy planned to remove Medley flash-weir, but in 1888 a boat-slide and side channel were built. Only in 1937 was the flash-weir removed.

Medley flash-weir probably began as a fish-weir for Rewley Abbey dedicated in 1281. Robert d'Oilly II founded Rewley for Augustinian Canons under their first prior Ralf, but Henry VIII dissolved it in 1539. Virtually all traces are now beneath railway sidings and urban sprawl. From cartographic evidence of 1578 it seems that Rewley Lock was very near where Medley flash-weir was sited. However, it was Rewley Middle Lock (location unknown) that had to be closed so that Castle Mill had sufficient water.

> *Manor Farm until the mid-1800s was Midley Farm from Mideley meaning middle island – between Thames streams. The lord of Wytham gave the farm to Godstow Nunnery in the 1140s. There were two corn-mills then, later used as fulling-mills and their tithes went to the nunnery. Farming was good here in the sixteenth century, but implementation of the Enclosure Act brought poverty.*

The Thames, which has now travelled more than 50 miles, still has double that to go before reaching the sea. It soon comes to Manor Farm and the narrows beneath Medley footbridge (SP498075), built in 1865 by public subscription when Henry Grant was sheriff at Oxford. The bridge, known as Rainbow Bridge, was made at Reading Ironworks and the ornate wrought iron has recently been refurbished. The long-established Bossom's boatyard nearby has expanded its activities over the decades with slipways for pleasure craft. In the early nineteenth century Bossom's were traders with four barges.

> *Oxford probably began as one of King Alfred's new burhs that were often sited to guard well-established river crossings founded in Roman times. Oxford was a royal estate of dwellings and religious foundations on the Mercia–Wessex border, which the river defined. From about 750 there was a religious community at St Frideswide's minster, which became a priory by 1122. Frideswide's contemporaries founded riverside minsters at Eynsham and Bampton. This emphasises the value of the river for travel and of the adjacent agricultural land. Oxford grew into a thriving medieval town. St Aldate's and St Frideswide's areas probably developed by 800, but northern Oxford had previously held an important position on the boundaries of three Iron Age tribes and there was a large defended site near Dorchester-on-Thames. A north–south route across the river grew in the tenth century and from the fourteenth century main routes from London to Bristol and South Wales passed through Oxford.*

> Oxford had a grid street plan and fortified walls from 912, though they are mostly lost now. Nevertheless, the line of later walls, a gateway and a bastion may be traced and it was the threat of Viking invasions in the mid-ninth century that led the inhabitants to protect their settlement against invaders who were over-wintering in England. More attacks came in 1009, when Thorkill 'the Tall' led a Danish army into the city and burned a great deal of it. King Aethelred asked the Danish army for peace in 1011. The Danes penetrated much of England south of the Thames, as far west as Wiltshire, and by 1013 Oxford surrendered to King Swein of Denmark.

A significant number of archaeological finds recovered from the Thames seem to relate to the Danish raids, with weapons being discovered in the river at Folly Bridge and at North Hinksey. From the same locations, pre-Roman bronze spearheads and daggers suggest earlier battles at the river-crossings. Some weapons may have been deliberately committed to the water, as the Thames is believed to have been a sacred river in the late Bronze Age and Iron Age.

Castle Mill Stream

Opposite Weir Cottage at Medley a path across Port Meadow follows the Castle Mill Stream – at one time described as the 'old navigation'. There is a bridge (on an 1876 map) at Sheriff's Crossing close to Walton Ford and a railway bridge at SP503073 over the stream. Castle Mill Stream then flows beside the Oxford Canal – an urban waterway frequented by joggers, dog-walkers and office-workers on lunch-breaks. Along its banks are numerous narrowboats on semi-permanent moorings.

Castle Mill Stream reaches Hythe Bridge (SP508064), built around 1200 by the monks of Osney Abbey, which was founded as an Augustinian priory in 1129 by a nephew of the city sheriff. The monks had at least four mills on Thames streams. Adjacent to Hythe Bridge fish-traps and baskets were made from osiers brought by barges during the seventeenth century. John le Pecher lived close by in the 1660s, and the nearby *wyre-ham* had become Wareham Yard by the early twentieth century.

When this area was a hive of industry, a floating chapel for boatmen was moored adjacent to Hythe Bridge in 1838, but the chapel sank in 1868 and was replaced by one on the riverbank. In 1767 Royal Assent was given for improvements to Hythe Bridge with a tollhouse. Later, a toll-bar was erected for the new turnpike

Medley old flash-weir and boat-slide, early 1900s (SP498073). (Courtesy of www.old-england.com)

Rainbow Bridge, Medley (SP498075). The Thames narrows beneath a footbridge built in 1865 by public subscription. It was made at Reading Ironworks and was recently refurbished.

routes via Botley Causeway to Swinford and Faringdon. Tolls ceased in 1868 and the tollhouse became the Gate House Inn (now the White House).

Castle Mill Stream soon comes to Pacey's Bridge (SP508063) on Park End Street, named New Road in the 1770s. The bridge was rebuilt in 1922 but with lower arches, which inhibited access by boat to Castle Mill. The area has since been developed with university accommodation blocks.

Quaking Bridge (SP509062) on Tidmarsh Lane was probably named as early as 1324, but a bridge existed in 1297. 'Tidmarsh' (maybe from *theod* meaning 'people' in Old English) possibly refers to a communally used lane. This area had eight breweries or malthouses in the nineteenth century. However, by the 1700s some of them had existed since the thirteenth century. These industries expanded when canals made coal transport more economical for the maltsters trading in the early eighteenth century.

In earlier times, Oxford had five mills. Records show Earl Algar had a mill within the city before 1066. Castle Mill ceased use in the 1920s but may well have predated the castle, which was begun around 1071 adjacent to the West Gate of the Saxon settlement. Castle Mill Stream flows at the foot of the old castle from which Matilda, when besieged by King Stephen, escaped in the winter of 1142 by being lowered on ropes from the castle before crossing the frozen waters of the braided Thames. The West Gate of the castle (later to become HM prison) had a timber bridge, in use by 1578, but about 1660 Castle Mill Bridge was described as 'a new bridge' – probably meaning 'newly-built bridge'.

The stone-built Swan Bridge (SP509061) has carried Paradise Street since at least the 1660s, though it has been widened since then. The street was named after the Paradise Orchard of the Dominican Friars.

> *The Dominican (Preaching) Friars, sometimes called Blackfriars, founded their first religious community at Oxford about 1221. In 1245 they moved to St Ebbe's Island, south of the Trill Mill Stream. The abbey became famous as a centre of learning, though any remains are now hidden beneath the Westgate Shopping Centre.*

Footbridges at SP508059 and SP509056 cross Castle Mill Stream before it flows under the A420 at Oxpens Road Bridge (SP509058), where it rejoins the Thames east of Osney Railway Bridge near Preacher's Pool.

Many corn mills at Oxford were driven by the Thames; grain and flour were often transported by barge. This mill on Castle Mill Stream was adjacent to Oxford Castle. (Courtesy of www.old-england.com)

Trill Mill Stream

About 200 yards south of the former castle, older maps show Trill Mill Stream, once controlled by *Aldewere*, which the king deemed 'a hindrance to navigation' (on Castle Mill Stream) in 1508. Aldewere may date from the time of Edward I, when it became a navigation flash-weir. Before the Conquest, Trill Mill Stream formed the southern boundary of the minsters of St Frideswide and St Aldate, and a 1912 *Oxford Chronicle* states that in ancient times it was a major Thames stream flowing within the city walls. Archaeology has shown that before the eighth century part of the stream was faced with timber, perhaps for a quay or mill. The name Trill may derive from *tirl* or *tyrl* (Old English), meaning a 'turning stream' for a horizontally set waterwheel. A religious community subject to Eynsham Abbey used the Saxon-built mill in 1005 and the Trill Mill Stream later drove the mill used by the Blackfriars of St Ebbe's.

In William II's time a major source of clean water for monastic buildings and the city was a *ductus aquae*, which came from the Corallian limestone hills west of Oxford. The conduit crossed the Thames, either by a bridge or over the riverbed, but the men of Seacourt destroyed it, for which they were fined 10*s* (50p). In 1288 fresh water for Blackfriars was brought by conduit from North Hinksey (Conduit House). This supply later became the basis of the city's fresh water. Early in the fifteenth century there was a butcher's quarter south of the Trill Mill Stream and in the mid-1600s the Trill flushed sewage from the town to an open sewer.

Today the Trill starts at a culvert through a rusty sluice gate (SP509059) and disappears underground, but it is possible to trace the course running eastwards, parallel with the old city southern walls. It passes beneath Norfolk Street (SP511059), which was Bridge Street until the early twentieth century. In the 1660s there was Preacher's Bridge (SP512059) on Old Greyfriars Street, which led to the neighbourhood of Blackfriars Abbey and the Greyfriars (Franciscans) church where many townsfolk were buried within its precincts.

The Trill Stream continues beneath Littlegate Street (SP512059) where Watergate, one of seven city gates, stood in 1244. A commemorative plaque a few paces up the slope towards the city marks the site of the Watergate. The stream continues beneath Rose Place and passes under St Aldates Street (SP514058), which was Fishstreet in 1369, but also once called South Gate. St Aldate's Street bridge is the site of a very ancient crossing place where a ford existed in 900 and may have served the nearby minster founded by St Frideswide.

Near the entrance to Christ Church Meadow at the north-eastern end of the War Memorial Gardens the Trill Mill Stream emerges from the culvert (SP515059), it then flows south along the eastern edge of the one-time St Ebbe's Island and passes beneath a footbridge (SP515056) at the back of Trill Court before its confluence with the Thames near the Head of the River Inn at Folly Bridge.

A sluggish (unnamed) stream from the Trill Stream crosses Christ Church Meadow to the river Cherwell. Some historians say this was Shirelake Ditch. However, we found a Shirelake Ditch earlier where it marked the county (or shire) boundary near Binsey in the seventeenth and eighteenth centuries. The Trill Stream once defined the Oxfordshire southern boundary at Littlegate Street and also on the eastern edge of St Ebbe's Island to a point on the southern edge of Christ Church Meadow. The boundary followed an arm of the Cherwell and then the 'so-called' Shirelake Ditch to a point opposite where Weirs Mill Stream leaves the Thames.

We must return to Medley footbridge and follow the Thames. We find tall trees flanking the navigation channel, which passes Fiddler's Island and soon comes to a water crossroads of Four Streams Bathing Place and Tumbling Bay on the Bulstake Stream; 'tumbling' derives from the late eighteenth century weir or lasher. In the recent past there was a foot-passenger ferry here (SP503066) and a tollhouse stood on a track beside the formerly navigable Bulstake Stream (1876 OS map). The bathing place was closed in 1990.

Opposite the Bulstake Stream is a narrow backwater, the Sheepwash Channel, which was widened and dredged to provide access via the Castle Mill Stream to the Oxford Canal at Isis Lock (SP505066), which was opened in 1796. The Sheepwash Channel flows under the towpath footbridge, a railway (SP504066) and a new bridge for a housing development.

Bulstake Stream and Hogacre Ditch

The Bulstake Stream is another diversion, but shown as 'River Thames or Isis' (1876 OS map), and whilst this was true because of the river's multiple channels, the Thames also powered the monks' mill at Osney. The Bulstake flows westwards from Four Streams Bathing Place and under Binsey Lane bridge where there is a hint of an earlier ford (SP498064). We might surmise that pilgrims visiting St Margaret's Well at Binsey used Binsey Lane and then perhaps crossed the Seacourt ford to reach Eynsham Abbey.

The Botley Stream that left the Seacourt Stream joins the Bulstake Stream in Bulstake Mead and then flows under Bulstake Bridge (SP497062), which was improved in the early 1660s during the construction of Botley Causeway. In the twelfth century, Eynsham Abbey farmed meadows nearby; might this account for a stone-carved niche set in the wall beside Bulstake Bridge? South of Botley Causeway the Bulstake Stream flows under a relatively new footbridge (SP497058) for a lane going beyond Ferry Hinksey Road to North Hinksey Village.

Fourteenth-century documents record that south of Port Meadow there was the 'ford-of-oxen' that gave Oxford its name and historians constantly search for its true location. In all probability there were many fording places. In the fourteenth century a ford was described as 'fit for cattle', and by 1467 (according to John Leland) there was a ferry. Another reference said a ford was 'west of Osney in Bullstake Mead'. If located west of Osney the ford must have crossed the Bulstake Stream and gone over the many areas of alluvial land between the streams.

The Bulstake flows parallel with the Thames then passes under an unremarkable iron-railing bridge (SP501055) over a track to South Hinksey before turning east at a spot where it is also called the Pot Stream. At this turn eastwards, the insignificant Hogacre Ditch runs off; it once defined the county boundary (1878 OS map). Hogacre Ditch probably joined with Eastwyke Ditch, which flows through land continually prone to flooding, and joins the Thames just north of Eastwyke Farm.

The Bulstake continues and rejoins the Thames near Osney Railway Bridge. Nearby is a horse bridge rebuilt in 1850 (SP507056) where the Thames towpath crossed over the Bulstake Stream. Adjacent to it is a stone memorial, erected by Oxford YMCA, to Edgar Wilson who at only twenty-one lost his own life while saving two drowning boys in June of 1889.

Back near Tumbling Bay Bathing Place, the present navigation channel, which even in the thirteenth century drove Osney Abbey Flour Mill, today passes the rear of terraced houses of Abbey Road (the site of Rewley Abbey). At Botley Causeway are the twin arches of old Osney Bridge (SP503062) that was begun around 1210 by monks to replace a wooden bridge. Osney's alternative spelling was Oseney, meaning *Osa's island*, and in the early thirteenth century the river here was known as the 'old river', because Castle Mill Stream took precedence. About 1467 Osney Bridge became known as *Hithebrigge* due to a nearby wharf; the name was corrupted to High Bridge during the eighteenth century. However, a city street-map of 1900 clearly shows Hythe Bridge crosses the Castle Mill Stream.

Parts of a seventeenth-century stone-arch bridge still exist. In 1885 one arch of Osney Bridge crumbled and a small child fell into the river; sadly the body was not recovered from the water until three years later. Osney Bridge arch was rebuilt, but at its opening in 1888 there were no celebrations as a mark of respect. The roadway at Osney Bridge was widened early in the twentieth century necessitating further alterations. Nevertheless, the adjacent iron bridge (SP503063) is an obstacle to larger cabin cruisers needing more than 7ft 6in (2.3m) headroom. This is because

Osney Bridge (SP503062) on Botley Causeway – a stone bridge begun c.1210 by monks to replace a wooden one. Part of the seventeenth-century stone arches still exists.

Standing oarsmen (an unusual sight) on the reach near Osney Railway Bridge.

raising the river level in the 1790s was to provide the water needed to drive the flourmill, and also sustain Osney Lock. Beneath the old stone bridge is a weir (a 'lasher' on older maps, referring to the rush of water) allowing a westward flowing channel, which bypasses Osney Lock.

After Osney Bridge the navigation channel soon reaches Osney Lock (SP503059), planned in 1787 and built by 1790 and now modernised with electrically operated gates. In 1227 the Preaching Friars at St Ebbe's wanted Osney Abbey weir removed, because they believed it affected the working of their Trill Mill. In 1793 Osney Weir toll was 2½d (1.25p) per ton. Adjacent to the lock, we find the site of Osney Abbey and the remains of the flourmill, with its own millstream in 1876 The brick-built hulk was burnt out by a devastating fire in 1947. The mill made gunpowder during the Civil War.

It is now only a short distance from Osney Lock to Folly Bridge. Improvements to the towpath were called for in 1811 when three towpath bridges were planned to replace hazardous fords over side streams, such as the Bulstake, when joining the Thames. The previous year authorities in Oxfordshire and Berkshire squabbled over who would pay for this repair work; this implies the actual position of the county boundary was still unclear.

The next Thames crossing is Osney Railway Bridge (SP507056), built in 1845 by Brunel and altered in 1898. The railways, though initially considered a luxury for passenger traffic, would ultimately bring about the demise of canal and river transport into and out of the city after 1852 when Botley Road railway station opened.

The river now enters a stretch known as Preachers Pool. A footbridge (SP509055), originally built in 1845 as part of Brunel's broad-gauge Great Western Railway, was rebuilt with steel-girders in 1898. Today, pedestrians and cyclists use the bridge.

Within a short distance there is another footbridge (SP512056) built in 1882 for a gas main but also used by a light railway to the gasworks; pedestrians and cyclists have used it since 1972. This bridge is in an area once busy with the medieval Friary wharves on the southern shore of the former St Ebbe's Island.

Oxford's medieval South Gate on St Aldate Street opened to the Thames and the multi-arch Folly Bridge (SP515055). This crossing place was in all probability first established by the Romans. Some historians think a Saxon ford was built as early as 871, but it was probably King Offa of Mercia who built a causewayed ford. Offa, who died in 796, had successfully won land from the West Saxons in a battle

At Preachers Pool, Oxford, a bridge (SP509055) built in 1845 as part of Brunel's broad-gauge Great Western Railway, was rebuilt with steel-girders in 1898. Today, pedestrians and cyclists use the footbridge.

near Benson in 779. It seems that a large part of Berkshire was in Mercian hands as late as 844 despite a general collapse of that kingdom in 825. A nineteenth-century map shows Folly Bridge south of the Oxfordshire–Berkshire boundary.

It is believed that the first 'Folly Bridge' was built around 1085 by Robert d'Oilly. However, recent archaeology revealed early medieval timbers of seventh- to tenth-century date. This implies that the crossing existed before the *burh* was established. The Constable of Oxford enlarged the bridge known as Grand Pont at that time and made a grant of pontage in 1369.

A 1½-mile causeway southwards crossed a floodplain described as Swynshill Field. At the end of the seventeenth century the causeway had a total of forty arches. When John Ogilby's surveyors visited Oxford in the 1670s they recorded 'marsh and sand both sides of the causeway', and the 1900 OS map clearly indicates that much land south of the city was liable to flooding. Today, numerous small arches are clearly visible below the road just south of Folly Bridge.

The fourteenth-century Oxford city fathers hired hermits to collect bridge tolls and to be responsible for repairs in order that this valuable asset to trade routes would survive. The present bridge of three arches is the result of reconstruction in 1779, but there were disputes between the turnpike trustees and the Thames Commissioners about who should pay for the work. In the age-old tradition tolls were levied and the tollhouse at the northern end of the bridge dates from around 1827. Tolls ceased in 1850 when it was reckoned the debt was paid. A weir was built beneath one arch of Folly Bridge early in the 1800s, but was soon in need of repair and removed in 1884.

About halfway across the Grand Pont there used to be a structure known as Bachelor's Tower, which in the thirteenth century was called Friar Bacon's Study, due to its use by the Franciscan philosopher and scientist. By about 1650 the tower had become a ruin and was known as The Folly. It was demolished in 1779 when the bridge was reconstructed. Beside the bridge, a more recent building dating from 1849 and known as Caudwell's Castle, has a number of white statues set in its Venetian-style façade.

The Thames now makes its way to Abingdon and the former towpath continues on the right bank. In the early 1700s the river here almost dried up due to a drought. The so-called Shirelake Ditch enters on the left bank of the river, where nearby there was once a ferry (SP523045).

King's boat-builders on Folly Bridge Island were taken over and expanded by John Salter in the 1850s. He later ran steamboats and established a landing stage where pleasure boats of every kind could be hired. Locals participated in the popular pastime of boating for pleasure, exchanging the less than pleasant air of the city for something purer in the countryside.

University life has over the years, from about 1815, included competitive rowing. The Boat Club, dating from 1839, holds races in Trinity term (eights) and in Hilary term (torpids), which are both great social events. The winner of the races becomes Head of the River. Adjacent to Folly Bridge is the Head of the River Inn, once a grain warehouse with a wharf in the early nineteenth century; its crane still stands on the terrace in front of the inn.

Oxford city enjoys a great many traditions, including wealth, which is scattered among the dreaming-spired churches, the museums, colleges and parks; a city 'whispering from her towers the last enchantment of the Middle Ages' (Matthew Arnold). During the mid-thirteenth century the university developed from its legendary founding by King Alfred. Balliol College was founded in 1263, William Merton endowed Merton College the next year, followed by others later that century. More colleges began over the decades; Exeter began in 1314 and was rebuilt during the Victorian Gothic revival era, which no doubt pleased its students Edward Burne-Jones and William Morris. In 1375 William Wykham, Bishop of Winchester founded New College, but this was the start of a much greater story.

Folly Bridge (SP515055). Oxford's medieval South Gate on St Aldate Street opened onto the multi-arch Folly Bridge – a crossing-place in all probability chosen by the Romans. It is believed the first bridge was built in around 1085 by Robert d'Oilly. However, archaeologists found early timbers of seventh- to tenth-century date, suggesting a crossing was already in use when the Saxon *burh* was established.

If you are interested in purchasing other books published by Tempus, or in case you have difficulty finding any Tempus books in your local bookshop, you can also place orders directly through our website

www.thehistorypress.co.uk